国家林业和草原局普通高等教育“十四五”规划教材

食品专业英语

（第二版）

陈忠军　陈　霞　主编

中国林業出版社

内 容 简 介

本教材共分为 7 章。其中，第 1 章介绍食品科技论文的特点、食品科技论文的结构及食品科技论文的翻译写作方法；第 2 章介绍食品化学及食品营养；第 3 章介绍食品微生物及发酵食品；第 4 章介绍食品添加剂及酶制剂；第 5 章介绍食品加工；第 6 章介绍食品质量与安全；第 7 章介绍食品新技术。除第 1 章外，其余各章包含课文和阅读材料。各章中每课课文后均有单词、注释，阅读材料部分和相应的练习是为了扩大学生的专业知识和词汇量。本书的最后附有食品专业词汇，以及食品专业相关信息、主要数据库和主要大学及网址，可供学生及食品行业相关人员查阅。

图书在版编目（CIP）数据

食品专业英语 / 陈忠军，陈霞主编．—2 版．—北京：中国林业出版社，2022.8（2025.7 重印）
国家林业和草原局普通高等教育“十四五”规划教材
ISBN 978-7-5219-1814-4

Ⅰ．①食…　Ⅱ．①陈…　②陈…　Ⅲ．①食品工业－英语－高等学校－教材　Ⅳ．①TS2

中国版本图书馆 CIP 数据核字（2022）第 149471 号

课件

中国林业出版社 · 教育分社

策划、责任编辑：高红岩　李树梅　　责任校对：苏　梅
电话：（010）83143554　　传真：（010）83143516

出版发行：中国林业出版社（100009　北京市西城区刘海胡同 7 号）
E-mail：jiaocaipublic@163.com　　电话：（010）83143500
http://www.forestry.gov.cn/lycb.html
印　　刷：北京中科印刷有限公司
版　　次：2016 年 6 月第一版（共印 2 次）
2022 年 8 月第二版
印　　次：2025 年 7 月第 2 次印刷
开　　本：787mm×1092mm　1/16
印　　张：16.75
字　　数：410 千字
定　　价：45.00 元

编 写 人 员

主　编　陈忠军　陈　霞

副主编　郑　艳　满都拉

编　者　（按姓氏拼音排序）

陈　霞（内蒙古农业大学）

陈忠军（内蒙古农业大学）

方海田（宁夏大学）

满都拉（内蒙古农业大学）

毛学英（中国农业大学）

孙子羽（内蒙古农业大学）

汪建明（天津科技大学）

赵　洁（内蒙古农业大学）

郑　艳（沈阳农业大学）

郑煜焱（沈阳农业大学）

第二版前言

《食品专业英语》（第二版）是在中国林业出版社的组织协调下，借鉴了多所大学的专业英语教学经验和目前食品行业的新技术，在第一版的基础上修订编写而成。在汲取相关教材及第一版教材内容的基础上，力求做到内容安排的系统性，除了科技论文写作外，还涉及食品化学、食品营养、食品微生物及发酵食品、食品添加剂和酶制剂、食品加工与贮藏、食品质量与安全及食品新技术等多方面内容。内容的取舍尽量做到重点突出，层次分明，注重结合实际，反映食品科技领域发展的新趋势、新技术，具有广泛的适用性。

本教材由内蒙古农业大学陈忠军、陈霞担任主编，沈阳农业大学郑艳、内蒙古农业大学满都拉担任副主编。参加编写的还有中国农业大学毛学英、天津科技大学汪建明、宁夏大学方海田、沈阳农业大学郑煜焱、内蒙古农业大学孙子羽、内蒙古农业大学赵洁，最后由陈忠军统稿。具体编写章节为：陈忠军编写第3章Lesson 1、2、3，Reading Material 1；陈霞和赵洁编写第1章；郑艳编写第6章Lesson 3，Reading Material 1、2；满都拉编写第4章；方海田编写第2章Lesson 1、2、3，Reading Material 1；毛学英编写第5章 Lesson 1、2，Reading Material 2、3；孙子羽编写第7章；汪建明编写第3章 Lesson 4，Reading Material 2、3；赵洁编写第3章Lesson 5，第5章Lesson 3、Reading Material 1，第6章Lesson 1、2；郑煜焱编写第2章Lesson 4、5，Reading Material 2、3；食品专业相关信息获取、主要数据库、主要学术机构、大学网址由陈霞收集整理，总词汇表由郑艳整理。

本教材的编写目标是培养学生掌握阅读和写作英语科技论文的基本技能，提高专业英语词汇量，提高专业英语文献的阅读和理解能力。教材第1章主要介绍了食品科技论文的特点、食品科技论文的结构及食品科技论文的翻译写作方法。其余各章包含课文和阅读材料两部分，每课课文后均有单词、注释，阅读材料部分和相应练习可供学生选择，以扩大学生的专业词汇量和英语实践应用能力。

本教材适用于普通高等院校食品科学与工程、食品质量与安全、发酵工程等专业的大学生，也可供从事食品加工、食品发酵、酒类酿造以及相应管理的人员作为参考。

由于编者水平和各方面条件限制，教材中难免存在许多不足之处，恳请读者提出宝贵意见。

编　者

2022年3月

第一版前言

《食品专业英语》是在中国林业出版社的组织协调下，借鉴了多所大学的专业英语教学经验编写而成。在汲取相关教材的基础上，力求做到内容安排的系统性，除了科技论文写作外，还涉及食品化学、食品营养、食品微生物及发酵食品、食品酶制剂和添加剂、食品加工过程及食品质量与安全等多方面内容。内容的取舍尽量做到重点突出，层次分明，注重结合实际，反映食品科技领域发展的新趋势、新技术，具有广泛的适用性。

本教材由内蒙古农业大学陈忠军担任主编，内蒙古农业大学陈霞、沈阳农业大学郑艳担任副主编。参加编写的还有中国农业大学毛学英、天津科技大学汪建明、齐鲁工业大学于海峰、宁夏大学农学院方海田、沈阳农业大学郑煜焱、陕西科技大学徐丹，最后由陈忠军统稿。具体编写章节为：陈霞编写第 1 章，方海田、郑煜焱编写第 2 章，陈忠军、汪建明编写第 3 章，于海峰编写第 4 章，毛学英、陈霞、陈忠军编写第 5 章，徐丹、郑艳编写第 6 章，食品专业相关信息获取、主要数据库、主要学术机构、大学网址由陈霞收集整理。

本教材的编写目标是培养学生掌握阅读和写作英语科技论文的基本技能，提高专业英语词汇量，提高专业英语文献的阅读和理解能力。教材第 1 章主要介绍了食品科技论文的特点、食品科技论文的结构及食品科技论文的翻译写作方法。其余各章包含课文和阅读材料两部分，内容均选自国外原版资料，每课课文后均有单词、注释，阅读材料部分和相应练习可供学生选择，以扩大学生的专业词汇量和英语实践应用能力。

本教材适用于普通高等院校食品科学与工程、食品质量与安全、发酵工程等专业的大学生、专科生，也可供从事食品加工、食品发酵、酒类酿造以及相应管理人员作为参考。

由于编者水平和各方面条件限制，教材中难免存在许多不足之处，恳请读者提出宝贵意见。

编　者

2015 年 10月

目 录

Chapter 1

Scientific Paper of Food Science and Technology

英语食品科技论文是食品科技发展及现代化建设的重要科技信息资源，是记录人类科技进步的历史性文件，是世界各国食品领域的科学家、相关专业技术人员、研究人员、学者等交流的重要手段。它是科学技术人员及其他研究人员在科学实验的基础上，对食品科学相关研究领域的现象或问题进行的科学分析、综合研究和阐述，是对这些现象和问题本质及其规律性进行揭示后撰写而成的。主要用于科学技术研究及其相关成果的描述，是科研成果的体现。因此，能够阅读和撰写英语食品科技论文，获取相关领域的最新研究成果，无疑是食品科学与工程及相关专业本科生应该掌握的基本技能。

Lesson 1　食品科技论文的特点

科技论文是在科学研究、科学实验的基础上，对自然科学和专业技术领域某些现象或某些问题进行研究，运用概念、推断、推理、证明或反驳等逻辑思维手段，分析、阐述、揭示这些现象和问题的本质及其规律性而撰写成的论文，是对创新性成果进行理论分析、实验验证以及科学总结，并通过公开发表或答辩的文章体例。食品科技论文有别于人文社科类论文和一般议论文，具有创新性、学术性、科学性、规范性、应用性、理论性、准确性和独创性等特点。

1．创新性

食品科技论文不是对前人研究的简单重复，其必须是作者本人研究的，在相关领域的理论、方法或实践上获得的新进展或新突破。科技论文要能体现与前人不同的新理念、新方法、新成果等，可丰富国内外相关领域知识文库，利于国内外学术同行引用。

2．学术性

学术性是科技论文的主要特征。它以学术成果为表述对象，以学术见解为核心，在科学实验的前提下阐述学术成果和学术见解，揭示食品领域相关内容发展、变化的客观规律，探索科技领域中的客观真理，推动科学技术的发展。科技论文是否有学术价值，是评价其质量的公认标准。学术性，即科技论文具有从实践中概括出来的对某一事物的理性认识的特性，其标准可概括为“新、深、实”三个字。

“新”，指在论点或方法上具有创新性，或赋予某一论题以新意。

“深”，指研究有一定的深度，能抓住问题的本质，对推动学科建设和相关研究发展有积极意义。

“实”，指所用的材料具有权威性和时间的贴近性，论据是经过组织和加工的，不是简单事实的罗列。

3．科学性

论文的内容必须客观、真实，定性和定量准确，不允许丝毫虚假，要经得起他人的重复和实践检验；论文的表达形式也要具有科学性，论述应该清楚明白，不能模棱两可，语言准确、规范。科学性是保证科技论文质量的最基本要求，其内涵通常可分解为真实性、准确性、可重复性、可比性和逻辑性。

（1）真实性

科技论文的内容必须真实，资料可靠。科研实验设计缜密，科研数据客观，要尊重科学事实，不得随意取舍客观数据或歪曲科学结论。

（2）准确性

科技论文表述的数据、引用的资料应准确无误，其结论和评价应准确反映客观事物及其规律。记录实验数据时要准确，在论文撰写过程中，严禁引用“二次文献”，应选择最恰当的词语和准确的单词，仔细推敲相近词在表述上的细微差别，力争把得到的研究成果及其意义准确表达出来。

（3）可重复性

在相同的条件下，读者如采用科技论文介绍的技术和方法，应能获得与论文相似的结果或结论。这就要求科研设计必须合理，在写作时要详细介绍必要的、关键的内容，尤其是科研作者自己创新或改进的技术和方法，以使读者可重复出同样或类似的结果，从而对科研成果进行推广，使其具备确定的应用价值、经济价值和社会价值。

（4）可比性

科技论文的结果可与其他已报道的相同或相近的课题结果进行比较，确定其先进性。这就需要设立对比观察，采用统计学的方法处理观察结果，增加其可行程度。

（5）逻辑性

科技论文必须脉络清晰、结构严谨，论证的展开应符合思维的客观规律，不能出现违背逻辑学原理和规律的错误。

4．规范性

规范性是指对科技论文的语言文字和表述形式方面的质量要求，一般应做到以下几点：

① 语言文字规范。

② 术语规范，尽量使用规范的专业词汇。

③ 计量单位规范。

④ 科技论文格式规范。不同的杂志对科技论文的格式有不同的要求，应严格按照杂志要求及相关标准构建论文的基本结构。

5．应用性

应用性是指科技论文在理论、方法或技术上的实际应用价值。衡量论文的应用性，可从以下三个方面入手：

① 论文是否从学科研究和社会实际生活中产生，是否反映了科学研究的新成果、新问题。

② 论据是否是从科研或社会调查中取得的第一手数据，还是东拼西凑、道听途说的“无本之末”。

③ 是否解决或回答了学科专业或社会发展中提出的迫切需要解决的问题，其答案对教学或科研是否有直接或间接的指导意义。

Lesson 2 食品科技论文的结构

食品科技论文一般具有统一的结构，这个格式在几百年前确定下来，并且被认为是科研工作者们交流的最好方法。统一的格式可使读者快速了解文章大概内容，并有选择地阅读全文。某些读者倾向于仅仅阅读题目，某些读者会阅读题目和摘要，而想要获得详细信息的读者可能会阅读全文。

食品科技论文一般分为研究型（包括研究报告、发现发明等）和综述型。

食品研究型科技论文主要包括研究报告型和发现发明型。研究报告型是针对某一科学课题进行调查与考察、实验与分析后，将得到的事物现象、实验数据等原始资料，经统计学分析整理后，结合已有理论做出的判断，得出的结论。而发现发明型论文则描述了新发现产品或技术的背景、应用前景、本质、特性及其变化所遵守的规律等；或阐述所发明产品、工艺、方法等的性能、特点、原理、使用条件等。研究型科技论文的格式通常包括：题目（Title），作者（Authors），摘要（Abstract），关键词（Key Words），引言（Introduction），材料与方法（Materials and Methods），结果（Results），讨论（Discussion），结论（Conclusion），致谢（Acknowledgements），参考文献（References），附录（Appendix）。但有些杂志的格式也会有所变化，所以在投稿前，应按照不同科技期刊的要求修改格式。

题目	清晰地描述研究内容
作者	撰写科技论文的相关人员
摘要	简单描述做了哪些实验，得到哪些结果以及主要结论
关键词	作为索引的工具确保文章能够被目标读者检索出来
引言	描述研究背景及主要科学问题
材料与方法	描述如何获得数据
结果	描述所发现的结果
讨论	对所获得的结果进行讨论
结论	作为整篇文章的总结，描述科研结果的科学意义
致谢	对资助研究成果顺利获得的课题、机构等表示感谢
参考文献	对所引用的文献进行整理和统计
附录	向读者提供补充数据

1．题目（Title）

论文题目是科技论文的必要组成部分。它要求用最简洁、准确的语言反映文章的特定内容，把论文的主题明白无误地告诉读者，使读者一看到题目就能了解文章的大致内容，进而决定是否进行摘要或全文的阅读。由于大多数读者只是浏览索引中科技论文的题目，只有少数人研读整篇论文。因此，在拟定题目时应慎重选择每一个字，力求做到长短适中、概括性强、重点突出，使读者一目了然。如果题目拟定不恰当，则很有可能被其目标读者忽略。

论文题目的拟定一般具备以下特征：

① 准确反映论文内容，既不过于空泛和一般化，也不过于烦琐，可清晰地反映文章的具体内容和特色，明确表明研究内容及创新点。

② 题目用语简练、明了，以最少的文字概括尽可能多的内容。当内容层次较多，难以简短化时，可采用主副题目相结合的方法，主副题目之间一般用冒号（：）隔开。如“Stability of Microbiota Facilitated by Host Immune Regulation: Informing Probiotic Strategies to Manage Amphibian Disease”。

2．作者（Authors）

一般学术性期刊要求将作者名字置于题目下方，并采用如下格式：

作者姓名

作者工作单位名称及地址，邮政编码

作者姓名中应仅包括对本研究做出贡献且对其数据和结论负责的人员，在起草论文时，要确定论文的署名，列为作者需得到本人同意，所有作者须对文章的最终版本做出确认。文章的第一作者为文章主要观点和主要结果的拥有者，是该研究成果第一手资料的具体操作者和掌握者，是论文的主要执笔者。通讯作者是可对论文所阐明的结果和观点给予最准确和全面的解释，承担完整法律责任的作者。通讯作者一般是课题的总负责人，或者是研究生的指导教师，是论文对外责任的承担者。在论文投稿、修改直至被接收发表过程中的一切联络工作一般都由通讯作者负责。一般用“*”在通讯作者名字上方标明，而后在杂志要求的相应位置（大多数为首页页脚）标明“* Corresponding author”以及通讯作者的详细联系方式（如通信地址、电话号码、传真号码、E-mail 地址等）。一般会根据对文章的贡献大小进行作者排序，有些期刊要求通讯作者排在首位，有些期刊要求通讯作者排在末尾。

在作者姓名的下方还应注明作者的工作单位全称、邮政编码、电子邮件地址或联系电话等。要求信息准确清楚，使读者能按所列信息顺利地与作者联系。一个作者应写一个地址，如作者地址相同则写一个即可；如地址不同，需在作者姓名处以上角标的方式标明，并在姓名下方按顺序注明。与中文联系地址的写法不同，英文地址的顺序为从小到大，即院系、所在单位、所在地信息（城市、省、邮编、国家）。如“College of Food Science and Engineering, Inner Mongolia Agricultural University, Hohhot, 010018, China”。

3．摘要（Abstract）

摘要是一段能够简要概括整个科技论文描述科学问题，采用研究方法，取得科学结果及相关结论的文字。如果说题目是关于文章内容的最简单的描述，摘要则是对论文内容准确、扼要的表达，不加解释和评论，就可使读者了解整篇科技论文的梗概。通常认为摘要是论文的“眼”，它能够给读者提供足够的细节，使读者确定是否需要通读全文。几乎所有公开发表的科技论文都有短小简洁的英文摘要，这是食品科技论文的一般要求。不仅英文科技论文如此，国内大多数中文科技论文也要求提供英文摘要，因为摘要可为情报文献检索数据库的建设和维护提供方便。一般要求摘要不得包含参考文献、图、表等。许多作者在撰写食品科技论文时，最后写摘要，使其可准确反映文章的主要内容。

摘要的内容一般包括以下五部分：研究背景（Background）、研究目的（Purpose）、研究方法（Researching Methods）、研究结果（Results/Findings）和结论、建议、展望（Conclusion，Recommendations，Prospects）。

（1）研究背景（Background）

一般在摘要的开头简洁地描述为什么进行该项研究、该研究的重要性、该领域的研究难点或本研究可能产生的影响。例如：

Polyphenols from the leaves of sea buckthorn（*Hippophae rhamnoides* L.）are nutritious and bioactive substances that can be used as nutritional supplements.

在某些科技刊物论文的英文摘要中，为使摘要更加精炼，研究背景这一部分经常被省略。

（2）研究目的（Purpose）

准确描述研究目的，简洁概括研究范围。例如：

To improve their stability and bioaccessibility *in vivo*, chemical extracts of sea buckthorn leaves were, for the first time, encapsulated using electrohydrodynamic technology.

（3）研究方法（Researching Methods）

简要介绍研究内容和实验过程，包括所用的原理、理论、条件、材料、工艺、设备、手段、程序等，是完成研究的必要手段。需要强调的是，摘要中的研究方法，不同于正文中的“Materials and Methods”，这里仅解释“What”，而不是“How”。例如：

The microcapsules were characterized using scanning electron microscopy, Fourier-transform infrared spectroscopy, and thermogravimetric analysis. The extract and microcapsules were evaluated for total phenols and flavonoids, total antioxidant activity, and their inhibitory effects on metabolic syndrome-related enzymes (α-glucosidase, α-amylase, and pancreatic lipase) under *in vitro* simulated digestion.

（4）研究结果（Results/Findings）

简要列出该研究所取得的主要结果，包括观察、实验、研究的结果、数据或发现，得到的性能及效果等。叙述应具体、准确。例如：

The results indicated that the extract was successfully encapsulated; encapsulation protected polyphenols and flavonoids from degradation and increase their bioaccessibility in the intestine. The antioxidant activity and the inhibition of metabolic syndrome-related enzymes were better reserved after encapsulation.

（5）结论、建议、展望（Conclusion，Recommendations，Prospects）

包括作者对结果的分析、研究、比较、评价、应用，提出的问题、建议、预测和展望等。例如：

Our findings will help in promoting the potential of sea buckthorn as a nutritional supplement and expanding its commercial use.

4．关键词（Key Words）

关键词是为满足文献标引或检索工作的需要而从论文中提取出的词或词组，一般为3～8个。关键词既可以作为文献检索或分类的标识，其本身又是论文主题的浓缩，选择合适的关键词可吸引目标读者阅读全文。读者可从中判断论文的主题、研究方向、方法等。关键词一般位于摘要之后、正文之前，通常为名词或名词短语，如使用缩略词，则应为公认和普遍使用的缩略语，否则应写出全称。关键词之间可以加逗号、分号或空格进行分隔，最后一个关键词后面一般不加任何标点符号。例如：

Keywords: Probiotics; Starters; Olive fermentation; Lactic acid bacteria

Key Words: bacteriocin, class Ⅱ, inducing peptide, two-component regulatory system

5．引言（Introduction）

引言是正文的第一部分，又称前言、绪论，是对研究工作的目的、范围、相关领域已有研究工作、知识空白、理论基础和分析、研究设想、研究方法、实验设计等的简要说明。要求作者能够清晰地陈述本研究要解决的问题、研究背景及其重要意义。同时，对已发表的相关研究进行描述，阐明本研究的创新点及重要性。有时还会简单描述论文的实验、假设、论点，概述实验设计或方法等。总的来说，引言一般解释两个问题，即“What”“So what”。

众所周知，在科技期刊发表的论文必须符合以下四个标准，即：

① 论文的内容必须符合杂志的出版范围。

② 论文的质量（包括采用的方法、结果、写作等）足够高。

③ 所得到的结果是有重要意义并且值得阅读的。

④ 论文中必须有能表现其创新性的结果。

引言则包含以上内容中的前三个部分，即是否符合期刊出版范围、创新性及重要性。一般来说，引言由研究背景和研究目的两部分组成。

（1）研究背景

简洁准确地描述已发表的相关研究内容，强调存在的问题及其重要性，使读者充分了解所研究或讨论课题的背景信息，读者可以看到一个关于所研究课题系统全面的综述。此处引用的参考文献一般要具有很强的相关性，其篇幅取决于杂志及目标读者。对于复杂课题的研究性论文，这部分内容的篇幅则较长。如果文章的目标读者是普通学者，在引言中会较为详细地介绍研究背景以使读者全面了解论文内容。而如果目标读者是对研究内容熟悉的专家，其重点则转为介绍为什么要进行该项研究及其重要性。例如：

Choline is commonly supplemented to dairy cows in the form of rumen-protected choline chloride since it was discovered that the choline molecule is extensively degraded by ruminal microorganisms. Feeding rumen-protected choline chloride aims to supply choline for intestinal absorption and therefore improve metabolic status in the cow (De Veth et al., 2016).

Studies in humans have shown that a greater intestinal supply and degradation of choline into trimethylamine is associated with greater relative abundance of Firmicutes and lower relative abundance of Bacteroidetes (Cho et al., 2017). These are also the 2 predominant phyla of bacteria in the rumen (Salfer et al., 2018) whose abundances are influenced by the concentration of neutral detergent fiber (NDF) in the diet. Similarly, greater relative abundance of Firmicutes and lower relative abundance of Bacteroidetes result from a reduction in dietary NDF concentration (Plaizier et al., 2017), which may therefore influence ruminal degradation of choline. Recently, Arce-Cordero et al. (2021) reported greater molar proportion of propionate and lower NDF digestibility in continuous culture when unprotected choline chloride was supplemented with 30% NDF diets; such effects were not observed with 40% NDF diets, indicating that utilization of choline by ruminal microorganisms is dependent on dietary NDF and may have an effect on propionate-synthesizing bacteria in the rumen. To our knowledge, the effects of unprotected choline on ruminal bacterial populations and their possible association with the type of diet have not been published.

（2）研究目的

这部分内容会告知读者本研究的目的及其与已发表研究的相关性，强调该研究的创新点、不同处。一般需在前言末尾用 1～2 段文字描述清楚。例如：

Our objective was to evaluate the effects of unprotected choline chloride on the ruminal microbiome at 2 dietary NDF concentrations. Based on our previous results (Arce-Cordero et al., 2021), we hypothesized that unprotected choline chloride would differentially affect ruminal bacterial populations depending on dietary NDF concentration.

6．材料与方法（Materials and Methods）

这部分是科技论文中的重要部分，提供了验证科学结果和结论的必要信息，清楚而准确地描述了该研究的实验过程及步骤，可使读者重复其研究内容并验证其结果。

“Materials”主要描述研究对象（如菌株、实验动物、组织等）、药品、培养基、仪器等。“Methods”则清楚地按顺序描述了实验设计、实验基本过程和步骤，以及实验条件。最后还会对数据的分析、重复次数、数据表示及统计方法等进行表述。根据食品科技论文类型和研究内容的不同，这部分可能有不同的写法，但一般都会根据情况列出相应的小标题。例如：

Milk Supply and Heat Treatment

Mid-lactation sheep milk (pH 6.56±0.01) was obtained from Neer Enterprises Limited (Carterton, New Zealand). The main composition of 3 batches of sheep milk was analyzed using a MilkoScan FT1 (FOSS) and by SDS-PAGE as described by Ye et al. (2016), and is shown in Table 1. A small amount of sodium azide (0.01%) was added to the unheated milk as a preservative. The whole sheep milk was skimmed at 3,000 × g for 15 min at 25℃ using a bench centrifuge (Heraeus Multifuge X3R; Thermo Fisher Scientific Inc.). The skimmed sheep milk (6 mL) was transferred into 10 mL sealable glass tubes and the well-sealed tubes were then heated at a range of temperatures (67.5-90℃) and times (0.5-30 min) with continuous rocking in a thermostatically controlled water bath. After heat treatment, the milk samples were immediately immersed in cold running water for cooling to room temperature. The heated milk samples were kept at room temperature for 6 h before further analyses.

Analysis of Protein Composition

Milks and the supernatants obtained from acid-precipitated and ultracentrifuged milk samples were analyzed by reversed-phase high performance liquid chromatography (HPLC) using a reversed-phase C18 column (Aeris Widepore 3.6 μm XB-C18 RP; Phenomenex) to determine the protein composition, as described by Bobe et al. (1998). The quantity of native whey proteins (*β*-LG and α-LA) in heated SSM was calculated by comparing the relative peak areas of the heated SSM with the original unheated SSM. The quantity of whey proteins in the ultracentrifugal supernatants was determined by comparing the relative peak areas of the supernatant fractions of the heated SSM with the original unheated SSM. All peak areas of these chromatograms were determined using peak integration algorithm LabSolutions software (Shimadzu Corporation).

Statistical Analysis

All experiments reported were fully triplicated on 3 batches of sheep milk, and the results are presented as the mean ± standard deviation. Although there were some variations between different batches, the same trends and relationships as reported here have been observed for all samples examined to date. The data were plotted using GraphPad Prism 8.4.0 (GraphPad Software). Statistical analysis was performed using 1-way and 2-way ANOVA and Tukey's multiple comparison test at a significance level of $P<0.05$.

上面摘录了几段科技论文中的有关材料与方法的内容。第一段介绍了实验对象来源、实验所需试剂、设备厂家及热处理方法等；第二段介绍了蛋白质测定方法和酪蛋白胶束微观结构观测的方法；第三段是数据统计分析方法。

7．结果（Results）

这部分客观地描述了所获得的研究结果及其意义。一般通过数据分析后的图或表描述原始数据，并由文字进行叙述以回答在引言部分提出的科学问题或假设。作者会根据“材料与方法”的顺序分小标题或段落对每一部分的结果进行描述。此外，作者还会对原始数据进行适当解释，以帮助读者理解所获得研究结果的意义或重要性。例如：

Table 1 shows the average slaughter values of the studied beef cattle. As expected, male cattle were statistically ($P<0.001$) significantly heavier than the female cattle at the end of the fattening period. Warm carcass mass was statistically ($P<0.001$) significantly higher in male cattle compared to group B.

The share of different tissues in three rib cut is shown in Table 2. Share of *M. longissimus dorsi* differed significantly between groups ($P<0.05$). The higher share of *M. longissimus dorsi* by 4.81% was recorded in male cattle compared to the group B. A statistically significant difference was found in the share of fat tissue ($P<0.01$), which was higher in young cattle of group B. Taking into consideration, the higher share of the muscle tissue and lower share of fat tissue, the advantage is on the side of young cattle of group A.

The chemical composition of *M. longissimus dorsi* is shown in Table 3. Statistically ($P<0.05$) significant difference was found in the water content that was higher in male cattle and lipid content, which was statistically (P<0.01) significantly higher in beef cattle the group B.

The sensory characteristics of cooked meat are shown in Table 4. Tenderness and flavor have been identified as the most important attributes that determine eating quality of Europeans. Tenderness is one of the major criteria that contributes most to eating satisfaction and consumers are willing to pay more for tender beef.

Based on the results of sensory evaluation of *M. longissimus dorsi*, softness/tenderness of meat in female cattle was statistically (P<0.001) significantly better. Differences in sensory characteristics can be attributed to a different content of intramuscular fat.

在上面段落中，分别对肉牛的屠宰性能、不同组织的份额、背最长肌化学组成、熟肉感官特征和里脊肉的感官评定五个方面进行分析和差异比较。通过计算 P 值确定样本间的差异显著性（P<0.05 为差异显著）。由以上范例我们可以看出，研究结果的描述应该以数据为依据，结合图表客观分析，语言描述要层次分明、重点突出。

8．讨论（Discussion）

这部分描述了所得到的研究结果的意义。通常直接、明确地将研究结果与引言部分提出的问题或假设联系，并与已经发表的相关研究成果进行比较，以强调本研究在相关领域的影响，并引出将要进行的研究内容。

讨论的主要内容包括：①对研究的主要目的或假设的回顾，并探讨是否得到预期结果；②对最重要结果的概述，并指出是否可能支持引言中的假设，与已发表的相关结果是否一致；③对结果提出说明、解释或猜测，并描述根据这些结果得出的结论或推论；④指出在研究过程中遇到的问题，分析其对研究结果的影响，并提出未来的研究内容或方向；⑤指出结果的理论意义和实际应用。例如：

There is strong evidence from large prospective cohort studies and meta-analyses that yogurt consumption is associated with a lower T2D risk as compared to the total dairy intake, leading to the proposal that the health benefits may be linked to the fermentation process, the lactic acid bacteria (LAB), and/or host-microbial mechanisms that are induced during yogurt consumption (Drouin-Chartier et al., 2019; Fernandez et al., 2017; Guo et al., 2017). In this study, we report that feeding yogurt corresponding to the equivalent of two servings per day, hence replacing 7.6% of the daily energy intake, prevents insulin resistance and hepatic steatosis in diet-induced obese mice. Consistent with one of our hypotheses, we showed that the beneficial impact of yogurt on insulin sensitivity is partly driven by changes in gut microbiota composition and function (HDCA), as validated by fecal material transplantation in GF mice. In addition, we discovered that three branched-chain hydroxy acids (BCHA), derived from LAB-driven metabolism of branched-chain amino acidsl (BCAA) during milk fermentation, are abundantly found in yogurt and other fermented milk products but not detectable in milk. BCHA plasma levels were positively associated with healthier metabolic parameters in obese mice and interestingly, these BCAA hydroxyl metabolites are also detected in metabolic tissues, even in non-yogurt fed mice. Remarkably, both systemic and tissue levels of BCHA are reduced in obese and insulin-resistant mice and their levels were partially maintained upon yogurt treatment, indicating that BCHA are metabolically regulated and not merely reflecting yogurt intake. We further show that BCHA can directly regulate glucose metabolism in insulin target tissues, as demonstrated by their ability to blunt glucose production and stimulate glucose uptake in liver and muscle cells, respectively. These data thus provide evidence that BCHA are cell-autonomous modulators of glucose metabolism and may be involved in the beneficial metabolic effects of yogurt treatment through their action on both liver and muscle tissues. Future studies will be needed to determine the mechanisms underlying our findings of reduced BCHA levels in obesity. It will be important to determine whether some of the bacterial taxa found to be altered by obesity and yogurt treatment express the enzymes required to metabolize BCAA to BCHA and whether this could contribute to the yogurt effect.

在上述讨论部分中，第一句介绍了主要研究方法和假设的回顾；第二句至第七句概述了主要研究结果，并对结果进行了说明与解释；第八句指出了通过这些结果得出的结论；第九句和第十句提出了未来研究方向和该研究的理论意义。

9．结论（Conclusion）

结论是整篇文章的总结，综合说明科研结果的科学意义。结论不是正文中各段小结的简单重复，应主要回答“What”，即研究出什么。完整、准确、精炼地指出以实验或考察中得到的现象、数据和阐述分析作为依据得到的观点。有的文章也会在结论部分提出当前研究的不足之处，并对研究前景和后续工作进行展望。例如：

It is well recognized that members of *Bifidobacterium* exert beneficial health effects on their host, however current knowledge of their diversity, distribution across the host phylogeny, and metabolic capability in non-human hosts, especially in wild animal populations, is limited. This research provides novel insights into the host-microbe evolutionary relationships and genomic features of *B. castoris* isolated from geographically distinct wild mouse populations. Our initial observations on strain-specific carbohydrate metabolism repertoires and the presence of *eps* genes require further investigation to understand how *Bifidobacterium* adapts, persists and interacts with the animal host and explain the functionality of mechanisms underlying bifidobacterial metabolic activity. This could be achieved through a combination of experimental and in silico methods, including additional isolation experiments and whole-genome sequencing, combined with community analyses based on metagenomic approaches, as well as carbohydrate metabolism and transcriptomics assays.

上述结论段中，第一句提出了该研究领域的局限性，第二句提出了该研究的主要发现和创新点，第三句和第四句对研究前景和后续工作进行展望。

结论部分一般加标题“Conclusion”或“Summary”，但也有的不加结论标题，而在Discussion中加以论述，得出结论。一般来说，这部分只有1～2小段长。

10．致谢（Acknowledgements）

致谢位于正文后，参考文献前。简单地对给予本研究帮助或支持的个人或机构表示感谢。常用的句式包括：

We would like to thank … for helpful comments and discussions (technical support).

The authors are grateful to … for providing…

This work was supported by …

11．参考文献（References）

参考文献位于致谢后，补充材料之前，注明引用已发表的科学成果的来源。所有不是来源于作者研究成果且不是众所周知的信息都会以参考文献的形式出现。参考文献的格式及排列由各期刊的要求所决定，一般包括作者姓名、题目、出版刊物名称、出版年份以及卷、期、页等。

一般来说，参考文献的具体编排顺序有以下两种：

（1）按作者姓氏首字母排序

正文中引用处用小括号注明第一作者的姓氏及出版年代。如果引用了由同一作者在同一年出版的多篇文献时，可在年代后用 a、b、c 区分。例如：

Arslain, K., Gustafson, C. R., Rose, D. J. (2021). The effect of health prompts on product consideration, attention to information, and choice in large, online product assortments: The case of fiber. *Food Quality and Preference*, 94: 104329.

Cheng, H., Hiro, Y., Hojo, T., Li, Y. (2018). Upgrading methane fermentation of food waste by using a hollow fiber type anaerobic membrane bioreactor. *Bioresource Technology*, 267: 386-394.

Dong, R., Yu, Q., Liao, W., Liu, S., He, Z., Hu, X., Chen, Y., Xie J., Nie, S., Xie, M. (2021). Composition of bound polyphenols from carrot dietary fiber and its in vivo and in vitro antioxidant activity. *Food Chemistry*, 339: 127879.

Jain, A., Soni, S., Verma, K. K. (2021a). Combined liquid phase microextraction and fiber-optics-based cuvetteless micro-spectrophotometry for sensitive determination of ammonia in water and food samples by the indophenol reaction. *Food Chemistry*, 340: 128156.

Jain, A., Jain, S., Barasker, S., Agrawal, A.(2021b). Predictors of discogenic pain in magnetic resonance imaging: a retrospective study of provocative discography performed by posterolateral approach.*The Korean journal of pain*, 34 (4): 447-453.

Oliver, A., Chase, A. B., Weihe, C., Orchanian, S. B., Whiteson, K. (2021). High-fiber, whole-food dietary intervention alters the human gut microbiome but not fecal short-chain fatty acids. *Msystems*, 6(2): e00115-21.

（2）按在文章中出现的顺序排序

正文中引用时只需标明其序号，不需列出作者姓氏及出版年代。例如：

[1] Schiano A N, Harwood W S, Gerard P D, et al. Consumer perception of the sustainability of dairy products and plant-based dairy alternatives[J]. Journal of Dairy Science, 2020, 103(12): 11228-11243.

[2] Dekker P J T, Koenders D, Bruins M J. Lactose-free dairy products: market developments, production, nutrition and health benefits[J]. Nutrients, 2019, 11(3): 551.

[3] Crump A, Jenkins K, Bethell E J, et al. Optimism and pasture access in dairy cows[J]. Scientific Reports, 2021, 11(1): 1-11.

[4] Crestani M S, Grassi T, Steemburgo T. Methods of nutritional assessment and functional capacity in the identification of unfavorable clinical outcomes in hospitalized patients with

cancer: a systematic review[J]. Nutrition Reviews, 2022, 80(4): 786-811.

[5] Naibaho J, Korzeniowska M. Brewers' spent grain in food systems: Processing and final products quality as a function of fiber modification treatment[J]. Journal of Food Science, 2021, 86(5): 1532-1551.

12. 附录（Appendix）

这部分不是每篇论文必备的，主要包括正文中没有罗列出的对理解全文有帮助的原始数据、复杂的计算、公式推导等。如果附录不止一个，可对其进行编号。

与研究型科技论文相比，综述型科技论文主要是对某一科学课题或技术领域在一定时期发展状况的回顾总结，对现状的分析评论，并对未来进行预测展望，提出建议，指出方向。虽不一定具有首创性，但要综合反映其在近阶段取得的进展及发展方向，具有指导性，对科学技术的发展可起到承前启后的作用。其一般格式通常包括：题目（Title）、作者（Authors）、摘要（Abstract）、关键词（Key Words）、正文（Main body）、参考文献（References）。除正文外，其余部分内容与上述研究型科技论文相似。综述的主体部分是对综述基本内容及不同方面引用的原始观点和材料所作的论述和介绍，其主要包括对论述课题的简要介绍、该课题的进展及其研究价值、对已发表研究成果的比较、结论。

Lesson 3 食品科技论文的翻译和写作

英语科技论文的语言不同于文学作品的语言，有其特有的语言风格，如语言简洁，大量使用被动语态，频繁使用名词化结构、介词结构等。英文科技论文是用来陈述食品科技相关领域所发现的新事物，或研究事物的新方法，具有科学性、理论性、学术性、准确性、创新性、独创性和规范性等特点。因此，在其撰写和翻译时应注重事实和逻辑，要求技术概念明确清楚，逻辑关系清晰突出，内容正确无误，数据准确精密，文字简洁明了，符合技术术语的表达习惯，体现其科学、准确、严谨等特征。

1. 食品科技论文的翻译技巧

（1）翻译中的改换

翻译中的改换包括词义的改换，词类、句子成分、修饰词的转换，各类从句间的相互转化，以及语态的改换等。

词义的转换是指在一个词所具有的原始意义的基础上，根据上下文和逻辑关系进一步加以引申，选择适当确切的目标词语来表达，避免生搬硬套，使译文更加通顺流畅。翻译时，如果完全生搬硬套字典提供的意思“对号入座”，会使译文生硬晦涩，含糊不清，甚至令人不知所云。因此，遇到一些无法直译的词或词组时，应根据上下文和原词的字面意思，做适当的改换。需要注意的是，词义的引申不得超出原始意义所允许的范围。例如：

Lactobacillus plantarum has the coding capacity for the utilization of many different sugars, uptake of peptides and formation of most amino acids.

译文：植物乳杆菌具备利用多种糖，摄取肽类物质，并形成多种氨基酸的特点。

其中，“coding capacity”不翻译为“编码容量”，而是引申为“具备……的特点”。

The factors which are likely to influence the infectivity of bacteriophages do not stop here.

译文：可能影响噬菌体侵染能力的因素并不止这些。

其中，“do not stop here”引申为“并不止这些”，而不是直译的“不停留在这里”。

Lactobacilli are widely used in a variety of food fermentation processes, where they contribute to the flavor and texture of final products.

译文：乳杆菌广泛用于食品发酵过程中，并可赋予终产品以良好的风味和质地。

其中，“contribute”译为“赋予”，而不是“贡献”。

词类、句子成分、修饰词等由于中英文的差异，翻译时有时也会发生改换。例如：

This yoghurt sample differs from the previous one by texture.

译文：这份酸乳样品与前一样品的区别在于质地。

其中，“differs”在翻译时由英语动词转换为汉语名称。

Both of the substances are not soluble in water.

译文：这两种物质均不溶于水。

其中，“soluble”在翻译时由英语形容词转换为汉语动词。

On the other hand, probiotic properties have been linked to *L. plantarum* strains.

译文：另一方面，植物乳杆菌还具备益生特性。

其中，英语主语“probiotic properties”在翻译时，由英文主语转换成补语。

语态在翻译时也会发生转换。英文被动句与汉语主动句互换的情况在科技文献中比较常见。例如，汉语中“样品采集自……”“分为……”“测定了……”“评价了……”“添加……”“建立了……”等，一般英语中均使用被动语态。根据英国利兹大学 John Swales 的统计，英语科技论文中的谓语至少 1/3 是被动语态。这主要是由于科技论文侧重的是叙事推理，强调客观准确，因此尽量使用第三人称叙述，采用被动语态。例如：

For the survey of rind diversity, a total of 362 cheese rind samples were collected across 137 different types of cheeses by scraping the rind surface with a sterile razor blade.

All data in figures or in text are presented as mean ± one SEM unless otherwise indicated.

An expert panel was convened in October 2013 by the International Scientific Association for Probiotics and Prebiotics (ISAPP) to discuss the field of probiotics.

（2）翻译中的增减

科技论文进行英汉互译时，往往会根据语言习惯，在译文中增加一些原文中无其形而有其意的词，或减去原文中某些在译文中属于多余的词。其目的是使翻译完整得体，语言通顺。有时为了避免重复，在科技论文撰写时，常常省略一些词，但在汉语译文中省略了这些词，就会使译文晦涩难懂，这时就需要增译。表示动作意义的名称有时可以表示具体概念，翻译时应在其后加“现象”“方式”“情况”“作用”“效应”“过程”等汉语名词，使译文具体化。例如：

The resistance of *Lactobacillus plantarum* to the heat treatment was higher than that of *Lactobacillus kefir.*

译为：植物乳杆菌对热处理的耐受力强于高加索酸奶乳杆菌。

Mater can be changed into energy, and energy into mater.

译为：物质可以转化为能，能也可以转化为物质。

Included, is a description of dehydration techniques and the concurrent changes that industrialized regions saw during their development.

译文：本文描述了干燥技术及其在工业化应用时进行的一些优化操作。

（3）长句的翻译

科技论文的语言表达客观准确，逻辑性强，结构严谨，为了更好地记录自然界的现象和科技界的动态，科技论文所采用句子往往偏长，结构复杂。例如，句子的修饰成分较多，可能包括多个从句或短语。长句翻译的前提是要正确分析和理解句子结构，通过对句子语法结构的分析，弄清主句、从句、分词短语等，再根据主从关系，找出长句的中心和各层意思，梳理好层与层之间的逻辑关系。在此基础上，根据汉语的表达方式和表达习惯进行翻译。为了使译文既能准确无误地传达原文的意义，又能符合汉语的表达习惯和方法，我们常采用以下几种译法：

① 顺译法：有些英语长句所叙述的一连串动作按发生的时间先后顺序安排，或按逻辑关系安排，与汉语的思维习惯和表达方式比较一致，此时可按原文顺序由前向后逐次翻译。例如：

Although a symbiotic relationship between *Streptococcus thermophilus* and *Lactobacillus delbrueckii* subsp. *bulgaricus* is generally assumed, not all strains are actually compatible, and growth imbalance in fermentations with mixed cultures may occur.

在上述例句中，“Although”引导让步状语从句，“and”引导并列主句。该句可按原文顺序译出，翻译时稍作修改即可。因此，该句可译为“尽管通常认为在嗜热链球菌和德氏乳杆菌保加利亚亚种间存在共生关系，但并不是所有菌株都是可以共存的，有些混合发酵剂在发酵时甚至会出现生长不平衡的现象”。

② 倒译法：在英文科技论文中，有时会出现表达顺序与汉语不同，甚至相反的现象。如先交代主要内容，然后再使用从句或分词补充细节，或者先交代结果，再讲明原因或条件等。翻译时，需从英文原文的后面译起，自后向前，逆着原文的顺序翻译。例如：

Food additives may be defined as “a class of substances that are used in food production, processing, storage and packaging to extend shelf-life of purpose and to meet the processing needs for improving quality, color, flavor and taste”.

在上述例句中，“that”引导英语从句修饰宾语“substances”，而“to”引导目的状语，“and”引导并列目的状语。该句包含三层意思，即“食品添加剂可定义为一类物质”“这类物质用于食品生产、加工、贮藏和包装中”“其目的是延长食品货架期或满足食品加工中改善食品质量、色泽、风味、质地的要求”。根据汉语习惯，该句可采用倒译法，从后往前翻译，译为“食品添加剂是为达到改善食品色、香、味，满足加工需要，或延长货架期等目的，而在食品生产、加工、贮藏和包装中使用的一类物质”。

③ 分译法：有时原句包含多层意思，而汉语则强调意合，表达同样的意思，通常用短句、分句等逐点交代，分层次展开。在准确理解的基础上，弄清各个短语或从句间的关系，拆分长句。在翻译这类句子时，一般需要采用化整为零的分译法，就是将原文中的从句或某一短语先译出来，并通过适当的概括性词语和其他语法手段，使前后句联系在一起。整个句子可译成若干个独立的句子，其顺序基本不变，保持前后的连贯性。例如：

Unfortunately, misuse of the term probiotic has also become a major issue, with many products exploiting the term without meeting the requisite criteria.

在该例句中，可以看出“with”引导了一个补语从句解释主句“misuse of the term probiotic has also become a major issue”，若翻译成补语从句，则会使整个句子中心偏移。因此，该句可以利用分译法进行翻译，译为：“然而，‘益生菌’的滥用也成为一个重要的问题，如有些产品虽然使用了‘益生菌’这个概念，但是却没有达到其所需的标准。”

④ 综合法：有些长句顺译或者倒译都感不便，分译也有困难，此时就应仔细推敲，或按时间先后，或按逻辑顺序，有顺有逆、有主有次地对全句进行综合处理，以使译文最大限度地符合汉语表达习惯。在翻译时，可以通过综合分析法找到句子的主体结构，分清主句和从句。例如：

Erkmen and Karata HP-treated pasteurised whole cow milk inoculated with *Staphylococcus aures* ATCC 27690 at pressures in the range 50-350 MPa for 4-12 min at 20℃。

该例句中共有三个动词，即“HP-treated”“pasteurised”“inoculated”。我们需要先找到谓语。从句子结构中，不难看出“HP-treated”为本句的谓语，其主句为“Erkmen and Karata HP-treated milk”。“pasteurised”“whole cow”“inoculated with *Staphylococcus aures* ATCC 27690”均修饰“milk”。“at pressures in the range 50-350 MPa for 4-12 min at 20℃”补充说明高压处理的条件。因此，该句可翻译为“Erkmen 和 Karata 对接种了金黄色葡萄球菌 ATCC 27690 的巴氏消毒全脂乳进行了高压处理，条件为 20℃，50～350 MPa 处理 4～12 min”。

在翻译英语科技论文时需要一定的技巧和方法，其方法并不是固定的。在实际翻译中应根据句子的具体结构、逻辑关系、特点，弄清句子的语法关系，选择合适的翻译方法，灵活处理，使译文达到以下要求：叙述逻辑清晰；技术内容准确；汉语表达通顺。因此，在翻译时要努力提高自身英语水平，注意学习翻译技巧，总结经验，准确通顺地表达原文内容，译出既忠实于原文内容，又符合汉语规范的句子来。

2. 食品科技论文的写作

Lesson 2 中，已经说明了食品科技论文的一般文体结构，下面以研究型科技论文为例，主要介绍题目、摘要、引言、材料与方法、结果、讨论、结论的写作方法。作者、关键词、致谢、参考文献、附录等只要遵循写作规范即可，具体见 Lesson 2。

题目（Title）

题目要清晰地反映文章的具体内容和特色，明确表明研究内容及创新点，力求简洁有效、重点突出。建议在拟定题目时，尽量省略一些不必要的单词，如“A study of...”“Investigations of...”“Observations on...”。在题目中也要尽量避免使用不常用的缩略词、术语、首词字母缩写、字符、代号、公式等。大多数英文题目都有关键词，为表达直接、清楚，引起读者注意，一般将表达核心内容的关键词作为题目开头，如“Influence of inulin and demineralised whey powder addition on the organic acid profiles of probiotic yoghurt”。

在拟定英文题名时，需注意以下几点：

① 科技论文的题目一般由名词或名词短语构成，即一个或几个名词加上其前置和(或)后置定语构成，如“Comparison of identification systems for psychrotrophic bacteria isolated from bovine milk”“A review on food safety and food hygiene studies in Ghana”。在必须使用动词的情况下，一般用分词或动名词形式，如“Improving internal communication between marketing and technology functions for successful new food product development”。少数情况下

可以用疑问句作题名，以引起读者的兴趣，如“Can we improve global food security? A socio-economic and political perspective”。

② 题名长度：不应过长，国外科技期刊一般对题目的字数有所限制，可在投稿前根据杂志要求对题目进行修改。但总的原则是题目应准确、简练、醒目。

③ 题目中的冠词：在早年，科技论文题目中的冠词用得较多，但近些年有简化的趋势，凡可不用冠词的均可不用。

④ 题目中的大小写：不同杂志对题目的格式有不同要求，一般有以下三种格式。

a. 每个单词的首字母大写，但冠词、介词、连词小写，如“Comprehensive Metabolomics to Evaluate the Impact of Industrial Processing on the Phytochemical Composition of Vegetable Purees”；

b. 题目第一个单词的首字母大写，其余字母均小写，如“Comprehensive metabolomics to evaluate the impact of industrial processing on the phytochemical composition of vegetable purees”；

c. 全部字母大写，如“COMPREHENSIVE METABOLOMICS TO EVALUATE THE IMPACT OF INDUSTRIAL PROCESSING ON THE PHYTOCHEMICAL COMPOSITION OF VEGETABLE PUREES”。

⑤ 题目中的缩略词语：已得到整个科技界或本行业科技人员公认的缩略词语，才可用于题目中，否则轻易不要使用。

摘要（Abstract）

摘要的写作并没有一成不变的格式，但由于摘要是对原始文献不加注释或评论的准确而简短的概括，是读者判断文章可读性的主要依据，因此在写摘要时，应尽量用最精炼的语言阐述研究方法、结果和结论，给读者一个清晰的思路，使读者准确获得原始文献的主要信息。同时，要突出论文的创新独到之处，吸引读者。

英文摘要多使用第三人称，如“the paper”“the article”“the research”“the results”。在需要使用第一人称时，通常用泛指的“We”，而不用“I”。英文摘要的语态应视具体情况而定。在不考虑动作的实施者、强调研究结果时采用简洁的被动语态，如“Likewise, it can be anticipated that food access and utilization will be affected indirectly via collateral effects on household and individual incomes, and food utilization could be impaired by loss of access to drinking water and damage to health.”。但由于主动语态较被动语态的表达更有助于文字清晰、简洁及表达有力，且更易阅读，因而目前大多数期刊都提倡使用主动语态，国际知名科技期刊 *Nature* 和 *Cell* 等尤其如此，如“Here we report that, in mice, relatively low concentrations of two commonly used emulsifiers, namely carboxymethylcellulose and polysorbate-80, induced low-grade inflammation and obesity/metabolic syndrome in wild-type hosts and promoted robust colitis in mice predisposed to this disorder.”。

英文摘要的时态以一般现在时、一般过去时为主。通常研究背景、目的、结论建议部分使用一般现在时，如“The present study aims to evaluate the probiotic potential of lactic acid bacteria (LAB) isolated from naturally fermented olives and select candidates to be used as probiotic starters for the improvement of the traditional fermentation process and the production of newly added value functional foods.”。一般过去时用于描述撰写论文前已完成的工作，如

采用的研究方法、取得的研究结果等，如“None of the strains inhibited the growth of the pathogens tested.”。

对于英文摘要的篇幅，不同的杂志要求不同。美国《化学文摘》和《医学文摘》规定在300词以内，学术会议一般要求500词以内。因此，在投稿前，应仔细阅读不同期刊的要求对摘要进行修改。

引言（Introduction）

如Lesson 2所述，引言是食品科技论文正文的第一部分，一般分为研究背景、研究目的。研究背景一般包含以下信息，即“What have other done?”“Provide evidence: supported by limited number of relevant references.”研究目的包括“Why undertake this research?”“How does it relate to what has already been written?”“What is so different or special about your research?”

食品科技论文常以过去为基点写研究背景，指出就这一课题进行过什么相关研究、有何结论、存在何问题等，一般采用过去时。在写引言时常会出现如引言内容杂乱无章，参考文献太多，过分评论已发表的研究成果，遗漏已发表的重要发现，研究目的含糊不清等问题。因此，在写作时应注意以下几点：

① 尽量准确、清楚且简洁地指出所探讨问题的本质和范围，对研究背景的阐述做到繁简适度。

② 应引用“最相关”的文献以指引读者。优选包括相关研究的经典、重要和最具说服力的文献，避免引用“二次文献”，或者不恰当地大量引用作者本人的文献。

③ 采取适当的方式强调作者在本次研究中最重要的发现或贡献，让读者顺着逻辑的演进阅读论文。

④ 解释或定义专业术语或缩写词，以帮助读者阅读论文。

⑤ 适当地使用第一人称，如“We”“Our”，以明确指示作者本人的工作，如用“We tried to explain…”来替代“This study is tried to explain…”。叙述前人工作欠缺以强调自己的创新点时，应慎重且留有余地，可采用类似如下的表达，如“To our knowledge, …”“Until now, there is little information available about…”。

⑥ 在描述曾进行过的研究、已发表结论、存在问题，以及本研究目的和基本内容时，一般采用过去时。而阐述有关现象或普遍事实时，可采用一般现在时。

⑦ 常用句型。

a．表示研究目的与意图：

Our task/job in/of this article/paper was to…

The purpose/goal of this study/research was to …

In this paper, we hope (intend, attempt) to...

In this work/research article, we aimed at…

b．表示研究意义：

… is important/valuable from the viewpoint that…

To understand...is important for further research on…

c．描述前人的研究或发现：

… reported in his study that…

In …, … et al. found/discovered/concluded/reported that…

d．描述需要解决的问题：

The problems to be solved are …

The largely unsolved problems are …

材料与方法（Materials and Methods）

这一部分主要用于说明实验的对象、条件、使用的材料、实验步骤或计算过程等。该部分主要描述的是“How”，即对过程的描述要完整具体、符合逻辑，使读者无需借助于其他参考文献便可了解论文所用方法的原理、技术和分析步骤、技术路线。相对于其他部分而言，方法是英文科技论文中相对容易撰写的部分。但需要注意的是，作者在此部分必须提供给读者足够的细节，使他们能够重复该项研究。在该部分，尽量用小标题有条理地书写，通常以实验材料的来源及获取等描述开始，然后逐项介绍所进行的实验。通过方法部分的介绍，应该能体现出该研究设计的严谨性，数据来源的可靠性，以及统计分析的正确性，并能让他人重复得出一致或相似的结论。一般使用过去时和被动语态，因为所叙述的是已经发生过的客观事实。但如果描述的内容为不受时间影响的事实时，则可采用一般现在时。具体要求如下：

① 对材料的描述应清楚、准确。材料描述中应清楚地指出研究对象（样品、菌株等）的数量、来源和准备方法。对于实验材料的名称，应采用国际同行所熟悉的通用名，尽量避免使用只有作者所在国家的人所熟悉的专业名称。

② 对方法的描述要详略得当、重点突出。应给出足够细节让同行进行实验的重复，避免混入有关结果或发现方面的内容。一般来说，如果你采用的方法是新的，则需要对其进行详细介绍。如果不是，则只需引用所采用的经典方法的名称，提供参考文献，并描述在哪里做了修改。

③ 力求语法正确、描述准确，需要采用简洁的语言。

在撰写该部分时，常根据实验进行的先后顺序使用相应的小标题。常包含的有：

a. Materials（材料）：

Strains（Name, Source, Number）

Equipments（Model，Manufacturer’s name, Address）

Reagent（Name, Manufacturer’s name, Address）

b. Methods（方法）：

Type of Experimental Design

Nutrients Analysis

Microbial Analysis

Physicochemical Analysis

Sensory Analysis

Toxicological Quality

Number of Replication

Statistical Analysis

结果（Results）

结果是科技论文的重要组成部分，应直接描述本项研究所取得的所有结果，并对结果进行定量或定性的分析，揭示有关的数据和资料，简洁、清楚、直观地说明事实。这是显示全文科学创新点的主要部分，因此需要逻辑清楚地描述结果。需注意的是，对实验或观察结果的表达要高度概括和提炼，不要简单地将实验记录数据或观察事实堆积到论文中，应突出具有科学意义和代表性的数据。另外，所描述的实验结果要真实，即使有些结果与理论不符，也不可省略，需描述出来，并在讨论中加以说明和解释。

一般在正文中用文字描述即可。如果数据较多可采用经数据处理后的图或表来完整表述。在插入图或表时，需对图或表进行排序，并拟定可描述其内容的图名或表名。需注意的是，图名位于图的下方，表名位于表的上方。需注明的内容则以小一号字的形式位于图或表的下方。在正文描述的过程中，切忌简单重复图表中的数据，应侧重于描述重要数据、相关趋势、意义及相关推论等。在文字说明中，如果提及图表，表格通常用“Table+序号”表示，如“… was shown in Table 1.”；各种类型的图通常用“Figure+序号”表示，如“From Figure 1, we can see that …”。“Table”可以缩写为“Tab.”，“Figure”可以缩写为“Fig.”，但必须做到前后统一，如用缩写，则都使用缩写。由于本部分叙述或总结的研究结果为过去的事实，因此通常采用过去时。而对研究结果进行说明或由其得出一般性推论时，可使用一般现在时。由于不同的动词表示研究结果的可靠性不同，同时也可表示作者不同的判断、推理和态度，因此选用动词时应慎重。用于文字说明的动词通常有 show，provide，present，summarize，illustrate，reveal，display，indicate，suggest 等。

另外，尽管在讨论部分会详细解释研究结果，但在结果部分也应对原始数据进行适当解释，这将有助于读者理解作者此次研究结果的意义或重要性。通常包含以下内容之一：

① 根据本人的研究结果做出推论，常见句型如：

These results suggested/recommended that…

As shown in Table 1, we can find that …

② 作者解释研究结果或说明产生研究结果的原因，常见句型如：

These findings are understandable because …

This circumstance may be caused by …

③ 作者对此次研究结果与其他研究者曾发现的结果或理论模型做比较，常用句型如：

These results agree well with the findings of…

These findings were in accordance with the data of …

These results were highly consistent with the findings of …

The data of our system was significantly higher than that of …

当评论的内容为对研究成果可能的证明时，句子的主要动词之前通常加上 may 或 can 等一般现在时的情态动词。常见句型如：

The results can be explained by …

A possible explanation for this is that …

This may have occurred because …

在撰写过程中，应尽量用准确、简洁、清楚的文字，避免使用冗长的词汇或句子介绍或解释图表，并在括号或文字中标明，如“Bloomy rind cheese, such as Brie and Camembert, are heavily inoculated with fungi to create a dense rind that is usually white in appearance (Figure 1A).”“The antibiotic resistance for the lactobacilli strains is shown in Table 4.”。

讨论（Discussion）

讨论是对论文研究结果的分析和深化，通过将本研究的结果与以往研究结果相结合，指出与以往研究的异同，阐述结果的意义，阐述作者的见解，讨论尚未定论之处。通过与同类研究结果的比较、分析，旁征博引，从不同角度全面提出作者的建议或评论。通过指出与同类研究结果的异同，提出本项研究的创新性。在充分明确事实的基础上，提出本项研究结果可能具有的广泛的科学意义、学术价值及应用前景。在撰写过程中，应尽量做到直接、明确，使读者明确为什么这篇文章值得引起重视。具体需注意以下几点：

① 在讨论部分可适当简要地回顾研究目的并概括主要结果，但不要罗列结果。对结果的解释要重点突出，精炼、清楚。

② 推论要符合逻辑，避免得出实验数据不足以支持的观点和结论。在论证时一定要注意结论和推论的逻辑性、科学性。在探讨实验结果的相互关系或科学意义时，不要试图获得可以解释一切的结论。

③ 在表达科学意义和实际应用效果时，应适当留有余地，尽量避免使用“must”等词汇，而选择“could”“might”等词汇表示论点的确定性程度。

④ 在回顾研究目的、与已发表结果进行比较时，一般采用一般过去时，而阐述由结果得出的具普遍有效的结论或推论时，可以使用一般现在时。常见句型如：

The results provided substantial evidence for …

These experimental results supported the original hypothesis that …

Our findings are in substantial agreement with those of …

The present results are consistent with those reported in …

These results appear to refute the original assumption.

The results given in Figure 1 support/validate …

It should be noted that this study has examined only …

This analysis has concentrated on …

The findings of this study are restricted to …

The limitations of this study are …

We should like to point out that we have not …

However, the findings do not imply …

需要注意的是，对于一篇科技论文，结果和讨论是核心部分，篇幅往往占到正文的 50% 以上。在结果和讨论不易明显分开的情况下，这两部分可以合起来撰写。

结论（Conclusion）

英文食品科技论文一般都有结论，是在理论和实验的基础上，通过逻辑推理出的结果描述，与引言相互呼应。结论是作者根据自己的研究结果并经分析讨论后所归纳出来的本项研究结果的精华和浓缩表达。但有些 SCI（Science Citation Index，科学引文索引）杂志并不要求将结论作为论文正文的单独一部分，而是将结论列在讨论之后，往往是讨论的最后一段，这就需要在写作时灵活掌握。

这部分的语言除了要严肃、连贯、精炼、确切外，还要具有高度的概括性。与其他部分不同，现在时（特别是一般现在时和现在完成时）和将来时在结论部分使用的频率很高。这是由于结论部分通常总结了研究者到目前为止做了哪些工作，得到什么结果，这些结果对现在来说有什么影响、意义和价值，能够应用于哪些方面，解决哪些问题等。常见句型如：

… has been studied in this article.

Through …, it has demonstrated that …

Overall, our study has revealed that …

用主动语态时，经常用 this paper，this investigation，this study，this survey，the results，the analysis 或 we 等作主语。常见句型如：

This study/investigation clearly demonstrates/discovers/reveals/represents/discusses …

This research has shown/proposed/described …

在撰写结论时，作者有时会通过使用情态动词对动词加以修饰，来表示作者不太肯定及慎重的态度，或用以缓和语气。常用的情态动词有 will，can，may，及其过去式 would，could，might 等，表示“可能”之意。常见句型如：

This research might provide some additional useful information to …

… may be a natural and necessary phenomenon.

Chapter 2

Food Chemistry and Nutrition

Lesson 1 Carbohydrates

Carbohydrates are a major class of naturally occurring organic compounds, which come by their name because they usually have, or approximate, the general formula $C_n(H_2O)_m$ with n equal to or greater than three. Among the well-known carbohydrates are various sugars, starches, and cellulose, all of which are important for the maintenance of life in both plants and animals. Although the structures of many carbohydrates appear to be quite complex, the chemistry of these substances usually involves only two functional groups *ketone* or *aldehyde carbonyls* and *alcohol hydroxyl* groups. An understanding of stereochemistry is particularly important to understanding the properties of carbohydrates. Configurational and conformational isomerism plays an important role.

Carbohydrates are part of a healthful diet. The acceptable macronutrient distribution ranges (AMDR) for carbohydrates is 45%-65% of total calories. Dietary fiber is composed of nondigestible carbohydrates and lignin intrinsic and intact in plants. Diets rich in dietary fiber have been shown to have a number of beneficial effects, including decreased risk of coronary heart disease and improvement in laxation. There is also interest in the potential relationship between diets containing fiber rich foods and lower risk of type Ⅱ diabetes. Sugars and starches supply energy to the body in the form of glucose, which is the only energy source for red blood cells and is the preferred energy source for the brain, central nervous system, placenta, and fetus. Sugars can be naturally present in foods (such as the fructose in fruit or the lactose in milk) or added to the food. Added sugars, also known as caloric sweeteners, are sugars and syrups that are added to foods at the table or during processing or preparation (such as high fructose corn syrup in sweetened beverages and baked products). Although the body's response to sugars does not depend on whether they are naturally present in a food or added to the food, added sugars supply calories but few or no nutrients. Consequently, it is important to choose carbohydrates wisely. Foods in the basic food groups that provide carbohydrates—fruits, vegetables, grains, and milk are important sources of many nutrients. Choosing plenty of these foods, within the context of a calorie-controlled diet, can promote health and reduce chronic disease risk. However, the greater the consumption of foods containing large amounts of added sugars, the more difficult it is to consume enough nutrients without gaining weight. Consumption of added sugars provides calories while providing little, if any, of the essential nutrients.

Classification and Occurrence of Carbohydrates

The simple sugars, or monosaccharides, are the building blocks of carbohydrate chemistry. They are polyhydroxy aldehydes or ketones with five, six, seven, or eight carbon atoms that are classified appropriately as pentoses, hexoses, heptoses, or octaoses, respectively. They can be designated by more specific names, such as aldohexose or ketohexose, to denote the kind of carbonyl compound they represent. For example, an aldopentose is a five-carbon sugar with an aldehyde carbonyl; a ketohexose is a six-carbon sugar with a ketone carbonyl:

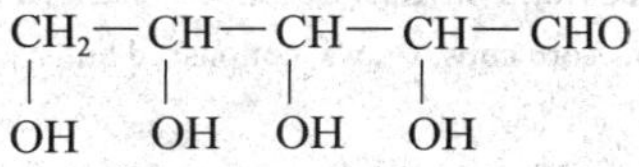

$$\underset{\text{OH}}{\underset{|}{CH_2}}-\underset{\text{OH}}{\underset{|}{CH}}-\underset{\text{OH}}{\underset{|}{CH}}-\underset{\text{OH}}{\underset{|}{CH}}-CHO$$

aldopentose

$$\underset{\text{OH}}{\underset{|}{CH_2}}-\underset{\text{OH}}{\underset{|}{CH}}-\underset{\text{OH}}{\underset{|}{CH}}-\underset{\text{OH}}{\underset{|}{CH}}-\underset{\text{O}}{\underset{\|}{C}}-\underset{\text{OH}}{\underset{|}{CH_2}}$$

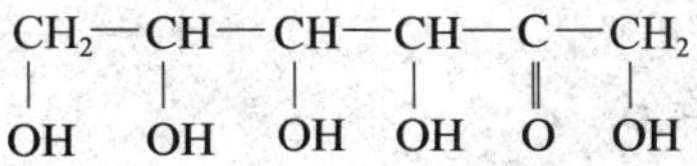

ketohexose

However, it is important to keep in mind that the carbonyl groups of sugars usually are combined with one of the hydroxyl groups in the same molecule to form a cyclic hemiacetal or hemiketal. These structures once were written as follows, and considerable stretch of the imagination is needed to recognize that they actually represent oxacycloalkane ring systems:

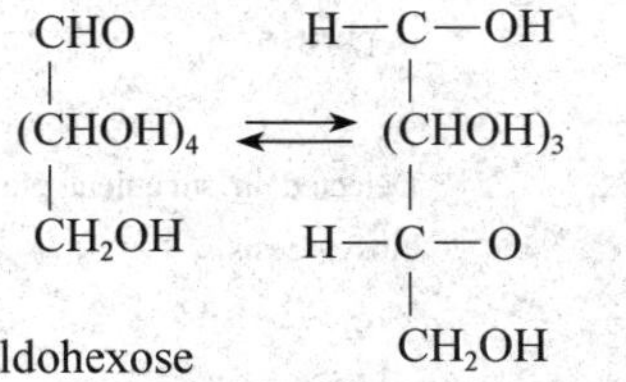

$$\begin{array}{c} CHO \\ | \\ (CHOH)_4 \\ | \\ CH_2OH \end{array} \rightleftharpoons \begin{array}{l} H-C-OH \\ \quad\ | \\ \quad (CHOH)_3 \\ \quad\ | \\ H-C-O \\ \quad\ | \\ \quad CH_2OH \end{array}$$

aldohexose (open-chain form) a 1,5-cyclic hemiacetal of aldohexose

The saccharides have long and awkward names by the International Union of Pure and Applied Chemistry (IUPAC) system, consequently a highly specialized nomenclature system has been developed for carbohydrates. Because this system (and those like it for other natural products) is unlikely to be replaced by more systematic names, you will find it necessary to memorize some names and structures. It will help you to remember the meaning of names such as aldopentose and ketohexose, and to learn the names and details of the structures of glucose, fructose, and ribose.

For the rest of the carbohydrates, the nonspecialist needs only to remember the kind of compounds that they are. The most abundant five-carbon sugars are L-arabinose, D-ribose, 2-deoxy-D-ribose, L and D-xylose, which all are aldopentoses. The common six-carbon sugars (hexoses) are D-glucose, D-fructose, D-galactose, and D-mannose. They all are aldohexoses, except D-fructose, which is a ketohexose. The occurrence and uses of the more important ketoses and aldoses are summarized in Table 2.1.

Table 2.1 Occurrence, physical properties, and uses of some natural sugars

Sugar	MP (℃)	$[\alpha]_D^{20-25}$ in H_2O[a]	Occurrence and use
Monosaccharides			
Trioses, $C_3H_6O_3$			
D-glyceraldehyde	syrup	+8.7	Intermediate in carbohydrate biosynthesis and metabolism
1,3-dihydroxy-2-propanone	81	—	As above
Tetroses, $C_4H_8O_4$			
D-erythrose	syrup	−14.5	As above
Pentoses, $C_5H_{10}O_5$			
L-arabinose	160	+105	Free in heartwood of coniferous trees; widely distributed in combined form as glycosides and polysaccharides
D-ribose	87	−23.7	Carbohydrate component of nucleic acids and coenzymes
D-xylose	145	+18.8	Called *wood sugar* because it is widely distributed in combined form in polysacchrides, such as in agricultural wastes as corn cobs, straws, cottonseed hulls
Hexoses, $C_6H_{12}O_6$			
D-glucose	146	+52.7	Free in blood, other body fluids, and in plants, abundant combined as polysaccharides
D-fluctose	102	−92.4	Free in fruit juices and honey; combined as in sucrose and plants polysaccharides
D-mannose	132	+14.6	Component of polysaccharides
D-galactose	167	+80.2	Called *brain sugar* because it is component of glycoproteins in brain and nerve tissues; also found in oligo and polysacchrides of plants
Heptoses, $C_7H_{14}O_7$			
Sedoheptulose	syrup	+8.2	Detected in succulent plants; an intermediate in carbohytrate biosynthesis
Oligosaccharides			
Disaccharides			
Sucrose	160~186	+66.5	Beat sugar and cane sugar (D-glucose+D-fructose)
Lactose	202	+52.6	Milk sugar and mammals (D-galactose+D-glucose)
Maltose	103	+130	Hydrolytic product of starch (D-glucose+D-glucose)
Trisaccharides			
Raffinose	78	+105	(D-glucose+D-fructose+D-galactose)
Polysaccharides			
Cellulose (poly D-glucose)			Occurs in all plants as a constituent of cell walls; as structural component of woody and fibrous plants
Starch (poly D-glucose)			As food reserves in animals (glycogen) and in plants

Note: a. Rotation at equilibrium concentration of anomers and of pyranose and furanose forms.

The recommended dietary fiber intake is 14 grams per 1,000 calories consumed. Initially, some Americans will find it challenging to achieve this level of intake. However, making fiber rich food choices more often will move people toward this goal and is likely to confer significant health benefits.

The majority of servings from the fruit group should come from whole fruit (fresh, frozen, canned, dried) rather than juice. Increasing the proportion of fruit that is eaten in the form of whole fruit rather than juice is desirable to increase fiber intake. However, inclusion of some juice, such as orange juice, can help meet recommended levels of potassium intake.

Legumes—such as dry beans and peas—are especially rich in fiber and should be consumed several times per week. They are considered part of both the vegetable group and the meat and beans group as they contain nutrients found in each of these food groups.

Key Recommendations

- Choose fiber rich fruits, vegetables, and whole grains often.
- Choose and prepare foods and beverages with little added sugars or caloric sweeteners, such as amounts suggested by the United States Department of Agriculture (USDA) Food Guide and the Dietary Approaches to Stop Hypertension (DASH) Eating Plan.
- Reduce the incidence of dental caries by practicing good oral hygiene and consuming sugar and starch containing foods and beverages less frequently.

Consuming at least half the recommended grain servings as whole grains is important, for all ages, at each calorie level, to meet the fiber recommendation. Consuming at least 3 ounce equivalents of whole grains per day can reduce the risk of coronary heart disease, may help with weight maintenance, and may lower risk for other chronic(*adj.* 慢性的) diseases. Thus, at lower calorie levels, adults should consume more than half (specifically, at least 3 ounce equivalents) of whole grains per day, by substituting whole grains for refined grains. Individuals who consume food or beverages high in added sugars tend to consume more calories than those who consume food or beverages low in added sugars; they also tend to consume lower amounts of micronutrients. Although more research is needed, available prospective studies show a positive association between the consumption of calorically sweetened beverages and weight gain. For this reason, decreased intake of such foods, especially beverages with caloric sweeteners, is recommended to reduce calorie intake and help achieve recommended nutrient intakes and weight control.

Total discretionary calories should not exceed the allowance for any given calorie level, as shown in the USDA Food Guide. The discretionary calorie allowance covers all calories from added sugars, alcohol, and the additional fat found in even moderate fat choices from the milk and meat group. For example, the 2,000 calorie pattern includes only about 267 discretionary calories. At 29% of calories from total fat (including 18 g of solid fat), if no alcohol is consumed, then only 8 teaspoons (32 g) of added sugars can be afforded. This is less than the amount in a typical 12 ounce calorically sweetened soft drink. If fat is decreased to 22% of calories, then 18 teaspoons (72 g) of added sugars is allowed. If fat is increased to 35% of calories, then no allowance remains for added sugars, even if alcohol is not consumed.

In some cases, small amounts of sugars added to nutrient dense foods, such as breakfast cereals and reduced fat milk products, may increase a person's intake of such foods by enhancing the palatability of these products, thus improving nutrient intake without contributing excessive calories. The major sources of added sugars are listed in Table 2.2.

Table 2.2 Major sources of added sugars (caloric sweeteners) in the American diet

Food categories	Contribution to added sugars intake (percent of total added sugars consumed)
Regular soft drinks	33.0%
Sugars and candy	16.1%
Cakes, cookies, pies	12.9%
Fruit drinks (fruitages and fruit punch)	9.7%
Dairy desserts and milk products(ice cream, sweetened yogurt, and sweetened milk)	8.6%
Other grains(cinnamon toast and honey-nut waffles)	5.8%

Note: Food groups that contribute more than 5% of the added sugars to the American diet in decreasing order.

Source: Guthrie and Morton, Journal of the American Dietetic Association, 2000.

The nutrition facts panel on the food label provides the amount of total sugars but does not list added sugars separately. People should examine the ingredient list to find out whether a food contains added sugars. The ingredient list is usually located under the nutrition facts panel or on the side of a food label. Ingredients are listed in order of predominance, by weight; that is, the ingredient with the greatest contribution to the product weight is listed first and the ingredient contributing the least amount is listed last. Table 2.3 lists ingredients that are included in the term "added sugars".

Table 2.3 Names for added sugars that appear on food labels

Brown sugar	Invert sugar	Fruit juice concentrates	Sucrose
Corn sweetener	Lactose	Glucose	Sugar
Corn syrup	Maltose	High-fructose corn syrup	Syrup
Dextrose	Malt syrup	Honey	
Fructose	Molasses	Raw sugar	

Note: Some of the names for added sugars that may be in processed foods and listed on the label ingredients list.

Sugars and starches contribute to dental caries by providing substrate for bacterial fermentation in the mouth. Thus, the frequency and duration of consumption of starches and sugars can be important factors because they increase exposure to cariogenic substrates. Drinking fluoridated water and/or using fluoride containing dental hygiene products help reduce the risk of dental caries. Most bottled water is not fluoridated. With the increase in consumption of bottled water, there is concern that Americans may not be getting enough fluoride for maintenance of oral health. A combined approach of reducing the frequency and duration of exposure to fermentable carbohydrate intake and optimizing oral hygiene practices, such as drinking fluoridated water and brushing and flossing teeth, is the most effective way to reduce incidence of dental caries.

Considerations for Specific Population Groups

Older Adults

Dietary fiber is important for laxation. Since constipation may affect up to 20% of people over 65 years of age, older adults should choose to consume foods rich in dietary fiber. Other causes of constipation among this age group may include drug interactions with laxation and lack of appropriate hydration.

Children

Carbohydrate intakes of children need special considerations with regard to obtaining sufficient amounts of fiber, avoiding excessive amounts of calories from added sugars, and preventing dental caries. Several cross-sectional surveys on U.S. children and adolescents have found inadequate dietary fiber intakes, which could be improved by increasing consumption of whole fruits, vegetables, and wholegrain products. Sugars can improve the palatability of foods and beverages that otherwise might not be consumed. This may explain why the consumption of sweetened dairy foods, beverages and presweetened cereals is positively associated with children's and adolescents' nutrient intake. However, beverages with caloric sweeteners, sugars and sweets, and other sweetened foods that provide little or no nutrients are negatively associated with diet quality and can contribute to excessive energy intakes, affirming the importance of reducing added sugar intake substantially from current levels. Most of the studies of preschool children suggest a positive association between sucrose consumption and dental caries, though other factors (particularly infrequent brushing or not using fluoridated toothpaste) are more predictive of caries outcome than sugar consumption.

New Words

isomerism ['aɪsəmərɪzəm] *n.* 异构；异构现象

laxation [læk'seɪʃən] *n.* 松弛；松懈；轻泻；轻泻剂

aldopentose ['ældəpentəʊs] *n.* 戊醛糖

cross-sectional [krɒs'sekʃənəl] *adj.* 分类排列的

discretionary [dɪ'skreʃənərɪ] *adj.* 任意的；自由决定的；酌情行事的；便宜行事的

cariogenic [kæraɪəʊ'dʒenɪk] *adj.* 生龋齿的

fluoridate ['flʊərɪdeɪt] *v.* 加氟；在饮水中加少量的氟（以防儿童蛀牙）

palatability ['pælətəbəlɪtɪ] *n.* 嗜食性；适口性；风味

ketose ['ki:təʊs] *n.* 酮糖

Notes

1) USDA: 美国农业部。
2) IUPAC: 国际纯粹与应用化学联合会。
3) Carbohydrates are a major class of naturally occurring organic compounds, which come by their name because they usually have, or approximate, the general formula $C_n(H_2O)_m$ with *n* equal to or greater than three.
   参考译文：碳水化合物是一类天然存在的有机化合物，名称的由来是因为它的分子式通常是 $C_n(H_2O)_m$，$n \geqslant 3$。
4) The simple sugars, or monosaccharides, are the building blocks of carbohydrate chemistry. They are polyhydroxy aldehydes or ketones with five, six, seven, or eight carbon atoms that are classified appropriately as pentoses, hexoses, heptoses, or octaoses, respectively.
   参考译文：单一的糖（或称为单糖）是构成碳水化合物化学分子式的基本单位。单糖是有五、六、七或八个碳原子的多羟基醛或酮，分为戊糖、己糖、庚糖或辛糖。
5) The most abundant five-carbon sugars are L-arabinose, D-ribose, 2-deoxy-D-ribose, L and D-xylose, which all are aldopentoses.
   参考译文：食品中最丰富的五碳糖是阿拉伯糖、D-核糖、2-脱氧-D-核糖和木糖，它们都是戊醛糖。
6) The common six-carbon sugars (hexoses) are D-glucose, D-fructose, D-galactose, and D-mannose. They all are aldohexoses, except D-fructose, which is a ketohexose.
   参考译文：常见的六碳糖除了果糖是酮糖，其他的（如葡萄糖、果糖、半乳糖、甘露糖等）都是醛糖。

Lesson 2 Protein and Amino Acid

Protein is the major structural component of all cells in the body. Proteins also function as enzymes, in membranes, as transport carriers, and as hormones; and their component amino acids serve as precursors for nucleic acids, hormones, vitamins, and other important molecules. The recommended dietary allowance (RDA) for both men and women is 0.80 g/(kg·d) of good quality protein and is based on careful analyses of available nitrogen balance studies (Table 2.4). For amino acids, isotopic tracer methods and linear regression analysis were used whenever possible to determine the requirements. The estimated average requirements for amino acids were used to develop amino acid scoring patterns for various age groups based on the recommended intake of dietary protein. The recommended protein digestibility corrected amino acid scoring pattern (PDCAAS) for proteins for children 1 year of age and older and all other age groups is as follows (in mg/g of protein): isoleucine, 25; leucine, 55; lysine, 51; methionine+cysteine (SAA), 25; phenylalanine+tyrosine, 47; threonine, 27; tryptophan, 7; valine, 32; and histidine, 18. While an upper range for total protein in the diet as a percent of total energy intake was set at no more than 35% to decrease risk of chronic disease, there were insufficient data to provide dose-response relationships to establish a tolerable upper intake level (UL) for total protein or for any of the amino acids. However, the absence of a UL means that caution is warranted in using any single amino acid at levels significantly above that normally found in food.

Table 2.4 Protein RDA as a percentage of energy RDA

Age(y)	Protein RDA(g/kg)	Protein RDA as a percentage of energy RDA (%)
0-0.5	2.2	8.0
0.5-1	1.6	6.5
1-3	1.2	4.9
4-6	1.1	5.3
7-10	1.0	5.6
11-14(males)	1.0	7.2
15-18(males)	0.9	7.9
19-24(males)	0.8	8.0
25-50(males)	0.8	8.7
51+(males)	0.8	11.0
11-14(females)	1.0	8.4
15-18(females)	0.8	8.0
19-24(females)	0.8	8.4
25-50(females)	0.8	9.1
51+(females)	0.8	10.5

Source: Eleanor Noss Whitney, understanding nutrition.

Definition

Protein

Compounds composed of carbon, hydrogen, oxygen, and nitrogen atoms, arranged into amino acids linked in a chain. Some amino acids also contain sulfur atoms. A biological macromolecule of molecular weight 5,000 to several millions, consisting of one or more polypeptide chains.

Amino Acids

Building blocks of protein; each contains an amino group, an acid group, a hydrogen atom, and a distinctive side group attached to a central carbon atom.

Essential Amino Acids

Amino acids that the body cannot synthesize in amounts sufficient to meet physiological needs. There are eight essential amino acids. They must be supplied by the diet.

Nonessential Amino Acids

The body can synthesize these amino acids for itself, if it is given nitrogen to form the amino group and fragments from carbohydrate and fat to form the rest of the structure.

Conditionally Essential Amino Acids

Sometimes a nonessential amino acid becomes essential under special circumstances. For example, the body normally make tyrosine (a nonessential amino acid) from essential amino acid phenylalanine. But if the diet fails to supply enough phenylalanine, or if the body cannot make the conversion for some reason (as happens in the inherited disease phenylketonuria), tyrosine then becomes conditionally essential.

Peptide Bond

A bond that connects the acid end of one amino acid with the amino end of another, forming a link in a protein chain.

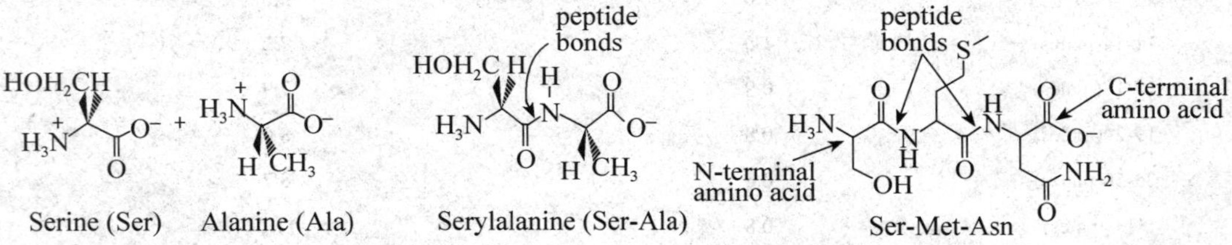

peptide bonds; N-terminal amino acid; C-terminal amino acid

Serine (Ser) Alanine (Ala) Serylalanine (Ser-Ala) Ser-Met-Asn

Peptide

The name given to a polymer of amino acids joined by peptide bonds; they are classified by the number of amino acids in the chain:

a. Dipeptide: a molecule containing two amino acids joined by a peptide bond.

b. Tripeptide: a molecule containing three amino acids joined by peptide bonds.

c. Oligopeptide: a molecule containing 4-9 amino acids joined by peptide bonds.

d. Polypeptide: a macromolecule containing 10 and more amino acids joined by peptide bonds. Polypeptide chains twist into different shapes of protein which enable protein to perform various functions in the body when growing or repairing/replacing tissue.

Chemistry of Proteins and Amino Acids

Protein

Protein is the major functional and structural component of all the cells of the body; for example, all enzymes, membrane carriers, blood transport molecules, the intracellular matrices, hair, fingernails, serum albumin, keratin, and collagen are proteins, as are many hormones and a large part of membranes. Moreover, the constituent amino acids of protein act as precursors of many coenzymes, hormones, nucleic acids, and other molecules essential for life. Thus, an adequate supply of dietary protein is essential to maintain cellular integrity and function, and for health and reproduction. The protein content in different foods is shown in Table 2.5.

Table 2.5 The protein content in foods

Foods	Content	Foods	Content
Milk	3.2%	Pork meat	12%
Yogurt	2.4%	Egg	12%
Bean sprouts	4.3%	Chicken	22%
Rice	6%-8%	Orange	0.8%

Proteins in both the diet and body are more complex and variable than the other energy sources, carbohydrates and fats. The defining characteristic of protein is its requisite amino (or imino) nitrogen group. The average content of nitrogen in dietary protein is about 16% by weight, so nitrogen metabolism is often considered to be synonymous with protein metabolism. Carbon, oxygen, and hydrogen are also abundant elements in proteins, and there is a smaller proportion of sulfur.

Proteins are macromolecules consisting of long chains of amino acid subunits. In the protein molecule, the amino acids are joined together by peptide bonds, which result from the elimination of water between the carboxyl group of one amino acid and the amino (or imino in the case of proline) group of the next in line. In biological systems, the chains formed might be anything from a few amino acid units (di-, tri-, or oligopeptide) to thousands of units long (polypeptide), corresponding to molecular weights ranging from hundreds to hundreds of thousands of Daltons. The sequence of amino acids in the chain is known as the primary structure. A critical feature of proteins is the complexity of their physical structures. Polypeptide chains do not exist as long straight chains, nor do they curl up into random shapes, but instead fold into a definite three dimensional structure. The chains of amino acids tend to coil into helices (secondary structure) due to hydrogen bonding between side chain residues, and sections of the helices may fold on each other due to hydrophobic interactions between nonpolar side chains and, in some proteins, to disulfide bonds so that the overall molecule might be globular or rod-like (tertiary structure). Their exact shape depends on their function and for some proteins, their interaction with other molecules (quaternary structure).

Many proteins are composed of several separate peptide chains held together by ionic or covalent links, an example being hemoglobin, in which each active unit consists of two pairs of dissimilar subunits (the α and β chains).

The most important aspect of a protein from a nutritional point of view is its amino acid composition, but the protein's structure may also influence its digestibility. Some proteins, such as keratin, are highly insoluble in water and hence are resistant to digestion, while highly glycosylated proteins, such as the intestinal mucins, are resistant to attack by the proteolytic enzymes of the intestine.

Amino Acids

The amino acids that are incorporated into mammalian protein are α-amino acids, with the exception of proline, which is an α-amino acid. This means that they have a carboxyl group, an amino nitrogen group, and a side chain attached to a central α-carbon. Functional differences among the amino acids lie in the structure of their side chains.

In addition to differences in size, these side groups carry different charges at physiological pH (e.g., nonpolar, uncharged but polar, negatively charged, positively charged); some groups are hydrophobic (e.g., branched chain and aromatic amino acids) and some hydrophilic (most others).

These side chains have an important bearing on the ways in which the higher orders of protein structure are stabilized and are intimate parts of many other aspects of protein function. Attractions between positive and negative charges pull different parts of the molecule together. Hydrophobic groups tend to cluster together in the center of globular proteins, while hydrophilic groups remain in contact with water on the periphery. The ease with which the sulfhydryl group in cysteine forms a disulfide bond with the sulfhydryl group of another cysteine in a polypeptide chain is an important factor in the stabilization of folded structures within the polypeptide and is a crucial element in the formation of inter-polypeptide bonds. The hydroxyl and amide groups of amino acids provide the sites for the attachment of the complex oligosaccharide side chains that are a feature of many mammalian proteins such as lactase, sucrase, and the mucins. Histidine and amino acids with the carboxyl side chains (glutamic acid and aspartic acid) are critical features in ion-binding proteins, such as the calcium-binding proteins (e.g., troponin C), critical for muscular contraction, and the iron-binding proteins (e.g., hemoglobin) responsible for oxygen transport.

Some amino acids in protein only achieve their final structure after their precursors have been incorporated into the polypeptide. Notable examples of such post-translational modifications are the hydroxyproline and hydroxylysine residues found in collagen (proline and lysine are converted to these after they have been incorporated into procollagen) and 3-methylhistidine present in actin and myosin. The former hydroxylated amino acids are critical parts of the cross-linking of collagen chains that lead to rigid and stable structures. The role of methylated histidine in contractile protein function is unknown.

Nutritional and Metabolic Classification of Amino Acids

Older views of the nutritional classification of amino acids categorized them into two groups: indispensable (essential) and dispensable (nonessential). The eight indispensable amino acids are those that have carbon skeletons that cannot be synthesized to meet body needs from simpler molecules in animals, and therefore must be provided in the diet. Although the classification of the indispensable amino acids and their assignment into a single category has been maintained in this report, the definition of dispensable amino acids has become blurred as more information on the intermediary metabolism and nutritional characteristics of these compounds has accumulated. Laidlaw and Kopple divided dispensable amino acids into two classes: truly dispensable and conditionally indispensable. Five of the amino acids are termed dispensable as they can be synthesized in the body from either other amino acids or other complex nitrogenous metabolites. In addition, six other amino acids, including cysteine and tyrosine, are conditionally indispensable as they are synthesized from other amino acids or their synthesis is limited under special pathophysiological conditions. This is even more of an issue in the neonate where it has been suggested that only alanine, aspartate, glutamate, serine, and probably asparagine are truly dietarily dispensable.

The term conditionally indispensable recognizes the fact that under most normal conditions the body can synthesize these amino acids to meet metabolic needs. However, there may be certain physiological circumstances: prematurity in the young infant where there is an inadequate rate at which cysteine can be produced from methionine; the newborn, where enzymes that are involved in quite complex synthetic pathways may be present in inadequate amounts as in the case of arginine, which results in a dietary requirement for this amino acid; or pathological states, such as severe catabolic stress in an adult, where the limited tissue capacity to produce glutamine to meet increased needs and to balance increased catabolic rates makes a dietary source of these amino acids required to achieve body nitrogen homeostasis. The cells of the small intestine become important sites of conditionally indispensable amino acid, synthesis, with some amino acids (e.g., glutamine and arginine) becoming nutritionally indispensable under circumstances of intestinal metabolic dysfunction. However, the quantitative requirement levels for conditionally indispensable amino acids have not been determined and these, presumably, vary greatly according to the specific condition.

There now appears to be a requirement for preformed α-amino nitrogen in the form of glutamate, alanine, or aspartate, for example. It was previously thought that, in addition to the indispensable amino acids, simple sources of nitrogen such as urea and diammonium citrate together with carbon sources would be sufficient to maintain nitrogen homeostasis. However, there are now good theoretical reasons to conclude that this is not likely in the human. The mixture of dispensable and conditionally indispensable amino acids as supplied by food proteins at adequate intakes of total nitrogen will assure that both the nitrogen and specific amino acid needs are met.

Protein and Amino Acid Metabolism

The ways by which different organisms utilize nitrogen source are different. Nitrogen fixation microorganism can directly utilize N_2 of air. Plants and some microorganisms can utilize, nitrite and ammonia, etc. Animals and peoples utilize amino acids to compound proteins and nucleic acids. But the processes of amino acid decomposition and synthesis of different organisms have many same points or similar point. Nitrogen metabolisms of different organisms have their own characteristics.

(1) Protein can be hydrolyzed into amino acid and oligopeptide by many proteases in gastric juice, pancreatic juice and intestinal juice of animals and peoples. Oligopeptide can be hydrolyzed into amino acid by oligopeptidase in mesentery cells. Amino acid enter cells is transmission through amino acid of membrane or glutathione. Then amino acid is transported into many organs (such as liver) or tissues by blood to construct proteins. Amino acid also can be decomposed to produce energy or transferred to be other active substances.

(2) Decomposition of amino acid can produce ammonia, α-keto acid, CO_2 and amine though deamination and decarboxylation. Amine can be decomposed into aldehyde and ammonia by amine oxidase. Aldehyde can oxidate to carboxylation. Ammonia is transferred to be urea in the liver of people and mammal, then urea is excreted through kidney. α-keto acid is decomposed through the way of carbohydrate metabolism or transfused to be carbohydrate and lipid. Ammonia is transferred to be urea through ornithine cycle. The key enzymes of this cycle are carbamyl phosphate synthetase I and argininosuccinate synthetase.

(3) Deamination of amino acid includes oxidative deamination, aminotransferation and combined deamination, Combined deamination is general. Only L-glutamic acid can carry out oxidative deamination. The main receptor of amino group in aminotransferation is α-ketoglutaric acid.

(4) One carbon group metabolism, transmethylation, sulfur amino acid metabolism and aromatic amino acid metabolism are relatively important. Aromatic amino acid metabolism is related to many genetic diseases of mankind.

Pathways of Amino Acid Metabolism

The exchange between body protein and the free amino acid pool is illustrated by the highly simplified scheme shown in Figure 2.1. Here, all the proteins in the tissues and circulation are grouped into one pool. Similarly, there is a second pool, consisting of the free amino acids dissolved in body fluids. The arrows into and out of the protein pool show the continual degradation and resynthesis of these macromolecules (i.e., protein turnover). The other major pathways that involve the free amino acid pool are the supply of amino acids by the gut from the absorbed amino acids derived from dietary proteins, the *de novo* synthesis in cells

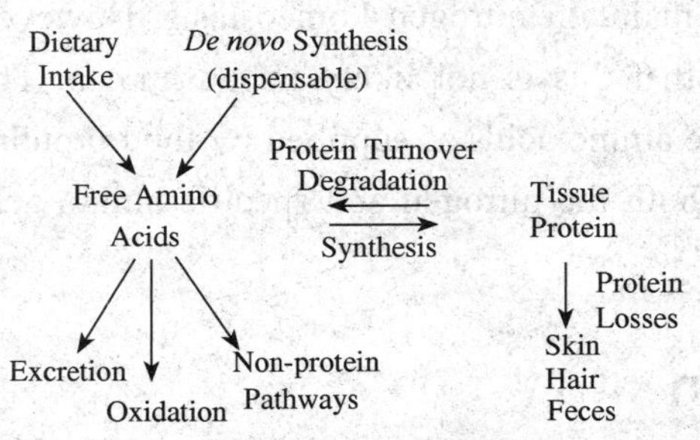

Figure 2.1 Exchange between body protein and free amino acid pools

(including those of the gut, which are a source of dispensable amino acids), and the loss of amino acids by oxidation, excretion, or conversion to other metabolites. Although this scheme represents protein metabolism in the human as a whole, with minor modifications it can also be used to represent protein metabolism in individual organs, or indeed the metabolism of a single protein.

Amino Acid Utilization for Growth

Dietary protein is not only needed for maintaining protein turnover and the synthesis of physiologically important products of amino acid metabolism but is, of course, laid down as new tissue. Studies in animals show that the composition of amino acids needed for growth is very similar to the composition of body protein. It is important to note, however, that the amino acid composition of human milk is not the same as that of body protein, and although the present recommendations for the dietary amino acids for infants provided in this report continue to be based on human milk as the standard, recent authors have cautioned that the composition of human milk proteins is not necessarily a definition of the biological amino acid requirements of the growing neonate.

Protein Synthesis

Amino acids are selected for protein synthesis by binding with transfer RNA (tRNA) in the cell cytoplasm. The information on the amino acid sequence of each individual protein is contained in the sequence of nucleotides in the messenger RNA (mRNA) molecules, which are synthesized in the nucleus from regions of DNA by the process of transcription. The mRNA molecules then interact with various tRNA molecules attached to specific amino acids in the cytoplasm to synthesize the specific protein by linking together individual amino acids; this process, known as translation, is regulated by amino acids (e.g., leucine), and hormones. Which specific proteins are expressed in any particular cell and the relative rates at which the different cellular proteins are synthesized, are determined by the relative abundances of the different mRNAs and the availability of specific tRNA-amino acid combinations, and hence by the rate of transcription and the stability of the messages.

From a nutritional and metabolic point of view, it is important to recognize that protein synthesis is a continuing process that takes place in most cells of the body. In a steady state, when neither net growth nor protein loss is occurring, protein synthesis is balanced by an equal amount of protein degradation. The major consequence of inadequate protein intakes, or diets low or lacking in specific indispensable amino acids relative to other amino acids (often termed limiting amino acids), is a shift in this balance so that rates of synthesis of some body proteins decrease while protein degradation continues, thus providing an endogenous source of those amino acids most in need.

Protein Degradation

The mechanism of intracellular protein degradation, by which protein is hydrolyzed to free amino acids, is more complex and is not as well characterized at the mechanistic level as that of synthesis.

A wide variety of different enzymes that are capable of splitting peptide bonds are present in cells. However, the bulk of cellular proteolysis seems to be shared between two multienzyme systems: the lysosomal and proteasomal systems. The lysosome is a membrane-enclosed vesicle inside the cell that contains a variety of proteolytic enzymes and operates mostly at acid pH. Volumes of the cytoplasm are engulfed (autophagy) and are then subjected to the action of the protease enzymes at high concentration. This system is thought to be relatively unselective in most cases, although it can also degrade specific intracellular proteins. The system is highly regulated by hormones such as insulin and glucocorticoids, and by amino acids.

The second system is the ATP-dependent ubiquitin-proteasome system, which is present in the cytoplasm. The first step is to join molecules of ubiquitin, a basic 76-amino acid peptide, to lysine residues in the target protein. Several enzymes are involved in this process, which selectively targets proteins for degradation by a second component, the proteasome. This is a very large complex of proteins, possessing a range of different proteolytic activities. The ubiquitin-proteasome system is highly selective, so can account for the wide range of degradation rates (half-lives ranging from minutes to days) observed for different proteins. It is thought to be particularly responsible for degrading abnormal or damaged proteins, along with regulatory proteins that typically are synthesized and degraded very rapidly.

Risk Characterization

Since there is no evidence that amino acids derived from usual or even high intakes of protein from food present any risk, attention was focused on intakes of the L-form of the amino acid found in dietary protein and amino acid supplements. Even from well-studied amino acids, adequate dose-response data from human or animal studies on which to base a UL were not available, but this does not mean that there is no potential for adverse effects resulting from high intakes of amino acids from dietary supplements. Since data on the adverse effects of high levels of amino acids intakes from dietary supplements are limited, caution may be warranted.

New Words

autophagy [ɔː 'tɑfədʒɪ] *n.* （细胞的）自我吞噬（作用）

cytoplasm ['saɪtəʊplæzəm] 细胞质

intestine [ɪn'testɪn]	*n.*	肠
	adj.	国内的；内部的
dose [dəʊs]	*n.*	剂量；药量；（药的）一服；一剂
	vt.	给（某人）服药

Notes

1）RDA：推荐膳食供给量，即营养学术权威机构向各国公众推荐的每日膳食中应含有的热能和营养素的量。根据人体对营养的需要，考虑了各项安全率（人体应激、个体差、食物烹调损失、人体消化吸收率及食物生产供应情况等）而制定的。

2）PDCAAS：蛋白质可消化性评分，一种评估可消化的蛋白质（转化成氨基酸）对发育生长期儿童营养价值的指标，由联合国粮食及农业组织（FAO）和世界卫生组织（WHO）推荐。如某蛋白质消化后含有人体全部必需氨基酸，其比例也符合人体需求，则该蛋白营养计分为1.00，浓缩的大豆蛋白质营养计分为0.99。

3）Essential amino acids：必需氨基酸，是人体（或其他脊椎动物）必不可少，而机体内又不能合成的，必须从食物中补充的氨基酸。对成人来讲必需氨基酸共有八种：赖氨酸、色氨酸、苯丙氨酸、甲硫氨酸、苏氨酸、异亮氨酸、亮氨酸和缬氨酸。如果饮食中经常缺少上述氨基酸，可影响健康。它对婴儿的成长起着重要的作用。组氨酸为婴幼儿（4岁以下）生长发育期间的必需氨基酸，精氨酸、胱氨酸、酪氨酸和牛磺酸为早产儿所必需。

4）Nonessential amino acids：非必需氨基酸，是人体（或其他脊椎动物）体内需要，但能自己合成的氨基酸。这类氨基酸不必由食物供给，对人来说非必需氨基酸为甘氨酸、丙氨酸、脯氨酸、酪氨酸、丝氨酸、半胱氨酸、天冬酰胺、谷氨酰胺、天冬氨酸和谷氨酸。非必需氨基酸的供给对于必需氨基酸的需要量是有影响的，例如，体内的酪氨酸（非必需氨基酸）可由苯丙氨酸转变而成，半胱氨酸可由蛋氨酸转变而成。

Lesson 3 Lipids

The chemical term lipid defines a class of distinct compounds; it also includes the “fats” as well as the sterols such as cholesterol. A better and simple way to class lipid compounds would be to state that these are substances including the fats, all oils or waxes that are not soluble in water but readily dissolves in alcohol. Like the other two primary organic compounds carbohydrates and proteins, the lipids are also made up of carbon, hydrogen and oxygen; they differ from carbohydrates mainly in that the proportion of oxygen in lipids tends to be much lower. All living cells require lipids as the lipids form an important structural part of the cell—besides performing other important roles inside the cell itself. All plant and animals cells contain lipids along with the other two primary organic compounds, namely proteins and carbohydrates. In fact, all cells are vital aggregations of these three organic compounds. Lipids include the important biomolecule cholesterol as well as the triglyceride—the principal storage fats in adipose tissues. The body readily stores lipids—these high energy biomolecules are used as a store of energy by all cells. The two main roles of lipids are their structural role in the cell architecture and their storage role as a source of fuel. In general, the term lipids cover the fatty acids, the neutral fats, as well as the waxes and steroids like the important hormone cortisone. Lipids are classified into many types, one important type being the compound lipids in which one kind of lipid is combined or chemically bound with another type of organic or inorganic partner, these class of lipids are important and include the combination with proteins called lipoproteins, the glycolipids which are a combination with carbohydrates as well as the phospholipids.

Some common properties shared by all classes of lipids are the fact that they are not soluble in water—they are non-polar molecules that are only soluble in some weakly polar or non-polar solvents including chloroform and alcohols. The formation of cellular structures is an important function of lipids, all kinds of cells in the body depend on lipids to stems to form sheet like membranes, that when combined with proteins forms the cell or plasma membrane—the basis of cellular life itself and by extension the most vital requirement for cellular organization in living systems. In both the simplest prokaryotic cells like bacteria and the more complex eukaryotic cells—animals and plants—the plasma membrane separates the cellular contents that form the living organism from the external fluid environment. Therefore, this membrane formed by lipids and proteins permits any cell to function as the basic unit of life on earth. In addition to the outer plasma membrane, the more complex eukaryotic cells of plants and animals have internal membranes surrounding small organelles inside the cell, like that surrounding the ER—endoplasmic reticulum, a complex of tubes and secretory granules—those around the nucleus—the controlling center of the cell, those around the mitochondrion—the powerhouse of the cell in animals, as well as those found around the chloroplast—the powerhouse of the cell in plants. These internal membranes which are also made from lipids and proteins add complexity to the cells of animals and plants and act to further compartmentalize the cell into specific units dedicated to unique roles. Inside the cell, one more important role that lipids play is as efficient storage molecules that can be oxidized to produce energy.

Membrane based lipids are of three major kinds, the phospholipids, the glycolipids, and the sterols such as cholesterol. A structure known as a lipid bilayer is rapidly formed by the chemical association of both phospholipids and glycolipids when they are in a fluid medium—this bilayer is the basis for the cell membrane. Once formed, the plasma membranes of cells act like two dimensional, semi fluid structures, which permit the embedded protein molecules in the membrane to constantly move freely in the lipid bilayer by lateral diffusion—therefore, cell membranes are dynamic and not fixed forms. In the simpler membranes of prokaryotic cells—bacteria—the cell membranes fluidity or dynamism is regulated by the varying of the number of double bonds as well as the fatty acid chains forming the lipid molecules of the membrane. Membrane fluidity and dynamism is a function of the quantity of the sterol lipid cholesterol in the cells of higher animals.

One very important aspect of the plasma membrane is its role as a selective filter that is responsible for controlling the entry of nutrients and other molecules into the cell—the membrane is the barrier to and regulator of the kind of molecules and compound that will enters the cell from the external environment. It may allow the entry of molecules useful to cellular processes and stop the entry of other substances which are dangerous or not required by the cell. The membrane is thus a conduit or a gate for all substances needed by the cell; it also expels the waste products of chemical metabolism and passes it out into the external environment for disposal. The chemical nature and the composition of membranes make them unique barriers, for example, membranes have a low permeability for ions and most polar molecules, these molecules are only allowed to pass through protein channels formed by integral membrane proteins—a special type of membrane protein—this allows them to be regulated. Most substances are taken into the cell by a process of diffusion, where substances move into the cell from an area outside the cell where they are found in high volumes to the area inside the cell where they are in low volume. For example, a scent diffuses about a room from one fixed point where it is sprayed, this is how most materials are taken in by the cell. If any substance is transported against the concentration gradient from an area of lower concentration outside the cell to an area of higher concentration inside the cell where its volume is already high, then some energy must be expended in a process called active transport—another way in which all cells take in substances they need.

The chemistry of food lipids is complicated because they are diverse types of compounds that undergo many interactions with other components of a food. Many important and well-understood chemical changes that occur in an isolated lipid may be modified by such factors as location of the lipid in a tissue system, the presence or absence of water, and the imposition of such stresses as heat or radiation. Metals, both in the free state as ions and as components of organometallic compounds, affect the chemistry of lipids, especially in oxidation reactions. Non-lipid components of a food may interact with lipids and this can produce changes in food quality.

The consumer and the processor of foods utilize substances from the nutrient group known as fats and oils. Fats and oils represent the most prevalent single category of a series of compounds known as lipids. The word "lipid" is defined in *Webster's Unabridged Dictionary* as "any of a group substance that in general are soluble in ether, chloroform, or other solvents for fats but are only sparingly soluble in water, that with proteins and carbohydrates constitute the principal structural components of living cells, and related and derived compounds, and sometimes steroids and carotenoids. This definition describes a broad group of substances that have some properties in common and have some compositional similarities. A classification of lipids proposed by Bloor contains the following elements, which are useful in distinguishing the many lipid substances:

(1) Simple lipids (neutral lipids) —esters of fatty acids with alcohols.

a. Fats: esters of fatty acids with glycerol.

b. Waxes: esters of fatty acids with alcohols other than glycerol.

(2) Compound lipids—compounds containing other groups in addition to an ester of a fatty acid with an alcohol.

a. Phospholipids (phosphatides): esters containing fatty acids, phosphoric acid and other groups usually containing nitrogen.

b. Cerebrosides (glycolipids): compounds containing fatty acids, a carbohydrate and a nitrogen moiety, but no phosphoric acid.

c. Other compound lipids: sphingolipids and sulfolipids.

(3) Derived lipids—substances derived from neutral lipids or compound lipids and having general properties of lipids.

a. Fatty acids.

b. Alcohols: usually normal chain higher alcohols and sterols.

c. Hydrocarbons.

Foods may contain any or all of these substances but those of greatest concern are the fats or glycosides and the phosphatides. The term "fats" applicable to all triglycerides regardless of whether they are normally nonliquid or liquid at ambient temperatures. Liquid fats are commonly referred to as oils. Such oils as soybean oil, cottonseed oil, and olive oil are of plant origin, lard and tallow are examples of nonliquid fats from animals, yet fat from the horse is liquid at ambient temperatures and is referred to as horse oil. Fats and oils also can be classed according to "group characteristics." Five well-recognized groups are the milk fat group, the lauric acid group, the oleic linoleic acid group, the linolenic acid group, and the animal depot fats group. The milk fat group pertains essentially to the milk of ruminants and especially to that of the cow, although in certain areas milk of the water buffalo or of sheep and goats may be prominent.

Milk fats are characterized by 30%-40% oleic, 25%-32% palmitic, and 10%-15% stearic acids. They generally have substantial amounts of C_4-C_{12} acids and are the only commonly used fats to contain butyric acid, which may be present in amounts from 3% to 15% depending on the source. Milk fat composition is particularly susceptible to variation as a consequence of the animal's diet. The lauric acid group is characterized by a high proportion (40%-50%) of lauric acid (C_{12}) and lesser amounts of C_8, C_{10}, C_{14}, C_{16}, and C_{18} acids. The unsaturated acid content is very low and this contributes to extremely good shelf life. These fats generally melt at low temperatures because of the short carbon chains present. The most widely used fats of this group are from the coconut, seeds of the oil palm, and the babassu or the coquilla nut. The oleic linoleic acid group, the largest and most varied group, contains only fats and oils of vegetable origin. These fats usually contain less than 20% saturated fatty acids, with oleic and linoleic acids being dominant. Such fats are commonly derived from seeds of cotton, corn, sesame, peanut, sunflower, and safflower and the seed coat or fruit pulp of the olive and the oil palm. Fats of the linolenic acid group contain substantial amounts of linolenic acid, although they also may contain high levels oleic and linoleic acid. The most important food oil of this group is that from the soybean. Others are wheat germ oil, hempseed oil, perilla oil, and linseed oil. The high linolenic acid content contributes to the drying oil characteristic, especially of linseed oil which contains up to 50% linolenic acid. The animal fat group consists mainly of lard from the pig and tallows from bovine and ovine sources. These are characterized by 30%-40% C_{16} and C_{18} saturated fatty acids and up to 60% oleic and linoleic acids. The melting points of these fats are relatively high, due partly to their contents of saturated fatty acid and to the types of glycerides present. With respect to the latter point, seed fats with as much as 60% saturated fatty acids often contain negligible quantities of trisaturated glycerides, whereas tallow with 55% saturated acids may contain up to 26% trisaturated glycerides. Differences in triglyceride composition affect physical properties, and this in turn greatly influences the use to which a given fat is put.

Role and Use of Lipids in Foods

Fats and oils are the most concentrated source of food energy. They provide 37.7 kJ of energy per gram which is approximately double the energy provided by proteins or carbohydrates. They are carriers of fat-soluble vitamins, and they contribute to food flavor and palatability as well as to the feeling satiety after eating. Lipids in the form of triglycerides, phospholipids, cholesterol, and cholesterol esters are important to the structure, composition, and permeability of membranes and cell walls. They perform a function of energy storage in seeds, fruits of plants and animals. Lipids are major component of adipose tissue, which serves as thermal insulation for the body, as protection against shock to internal organs, and as a contributor to body shape. Fats and oils are used as frying fats or cooling oils where their role is to provide a controlled heat-exchange medium as well as to contribute to color and flavor. As shortenings, they impart a "short" or tender quality to baked goods through a combination of lubrication and an ability to alter interaction among other constituents. As salad oils, they contribute to mouth feel and as carrier for flavors, and when emulsified with other ingredients they perform the same functions in the form of viscous pourable dressing or semisolid fatty foods known as mayonnaise or salad dressing. Margarines are used both for baking and cooking and as table spreads. Specially selected or manufactured fats are useful in confections, especially as enrobing or coating agents. These fats must have a short-melting rang at body temperature. Other fatty materials, such as the mono-and diglycerides, and certain phospholipids, such as lecithin, have useful roles as emulsifiers. Mono-and diglycerides contribute to shortening performance and act as staling inhibitors in bakery products. Lecithin is used as a mold release agent in confections, to control fat bloom in chocolate candied, and as an anti-spattering agent in cooking margarines. Fats and oils are available in a variety of forms. Butter, cooking oils margarines, salad oils, and shortenings are essentially all-lipid forms. Salad dressings and mayonnaise are composed of high proportions of fats or oils. Ingested fats and oils include not only those from obvious source but also those from invisible fat sources, such as cereals, cheese, eggs, fish, fruits, legumes, meat, milk, nuts and vegetables. This latter group constitutes approximately 60% of the dietary fat. Salad and cooking oils are prepared from cottonseed oil, soybean oil, corn oil, peanut oil, safflower oil, olive oil, or sunflower seed oil. These oils are usually refined, bleached, and deodorized. Some oils may be lightly hydrogenated to provide special properties and to enhance flavor stability. Margarines, used mostly as table spreads and to some extent as cooking fats, are prepared by blending suitably prepared fats and oils with other ingredients, such as milk solids, salt, flavoring materials, and vitamins A and D. The fat content must be at least 80%. Vegetable oils are used predominantly for manufacture of margarine although some animal fats are used. The fats may be single hydrogenated fats, mixtures of hydrogenated fats, or blends of hydrogenated fats and unhydrogenated oil. Special margarines are prepared in response to medical research, which implies a possible superiority for these types of margarines, especially for persons prone to atherosclerotic conditions. Commercial shortenings are semisolid plastic fats made with or without emulsifiers. Plasticity, or ability to be worked, is a major feature distinguishing these from other fats. Original shortenings consisted of lard or tallow, but hydrogenated vegetable oils and various combinations of fats are used to build in specific properties desired for baking. Cottonseed oil, soybean oil, tallow, and lard are the principal fats used in shortenings. However, no natural fat possesses all of the desired characteristics. Butter, obtained by churning cream, is a water-in-oil emulsion containing 80%-81% milk fat, which is present in plastic form. Other constituents in small amounts include casein, lactose, phosphatides, cholesterol, calcium salts, and usually 1%-3% sodium chloride. Varying but small amounts of vitamin A, E and D also are present, along with flavor bodies consisting of diacetyl, lactones, and butyric and lactic acids. Cocoa butter, derived from the cocoa bean, is a fat preferred for confectionary uses. It is usually in insufficient supply and is costly, so that many efforts have been made to substitute for it or to find suitable extenders.

New Words

well-understood *adj.* 确定的
imposition [ɪmpə'zɪʃən] *n.* 强迫接受；征税；欺骗
radiation [ˌreɪdɪ'eɪʃən] *n.* 辐射；发光；放射物
organometallic [ˌɔːgənəʊmɪ'tælɪk] *adj.* 有机金属的
oxidation [ˌɒksɪ'deɪʃən] *n.* 氧化
category ['kætəgərɪ] *n.* 种类；分类；范畴
unabridged [ˌʌnə'brɪdʒd] *adj.* 完整的；未经删节的；足本的
ether ['iːθə] *n.* 乙醚
chloroform ['klɒrəfɔːm] *n.* 三氯甲烷（氯仿）
vt. 用氯仿麻醉
solvent ['sɒlvənt] *adj.* 有溶解力的

	n.	溶剂；解决方法
steroid ['sterɔɪd]	*n.*	类固醇；甾族化合物
carotenoid [kə'rɒtənɔɪd]	*n.*	类胡萝卜素
neutral ['nju:trəl]	*adj.*	中立的；中性的
waxes ['wæksɪz]	*n.*	蜡；蜡状物
	vt.	给…上蜡
phospholipid [ˌfɒsfə'lɪpɪd]	*n.*	磷脂
cerebrosides[serɪb'rəʊsaɪdz]	*n.*	脑苷脂
sphingolipids [ˌsfɪngəʊ'lɪpɪdz]	*n.*	鞘脂类
triglyceride [traɪ'glɪsəraɪd]	*n.*	甘油三酯
ambient ['æmbɪənt]	*adj.*	周围的；外界的；环绕的
	n.	周围环境
soybean ['sɔɪbi:n]	*n.*	大豆；黄豆
lard [lɑ:d]	*vt.*	润色；点缀；涂加猪油
	n.	猪油
tallow ['tæləʊ]	*vt.*	涂脂油于；用油脂弄脏
	n.	牛脂；兽脂；动物脂油
well-recognized	*adj.*	公认的
lauric['lɒrɪc] acid	*n.*	月桂酸
linoleic [lɪ'nəʊli:k] acid	*n.*	亚油酸
depot ['depəʊ]	*vt.*	把…存放在储藏处
pertain [pə'teɪn]	*vi.*	属于；关于；适合
water buffalo ['bʌfələʊ]	*n.*	水牛
oleic [əʊ'li:ɪk]	*adj.*	油的；油酸的
palmitic [pæl'mɪtɪk]	*adj.*	来自棕榈的
stearic [stɪ'ærɪk]	*adj.*	硬脂的；硬脂酸的；似硬脂的；十八酸的
butyric [bju: 'tɪrɪk] acid	*n.*	丁酸
susceptible [sə'septəbl]	*adj.*	易受影响的；易感动的
unsaturated [ʌn'sætʃəˌreɪtɪd]	*adj.*	不饱和的
coconut ['kəʊkənʌt]	*n.*	椰子；椰子肉
palm [pɑ:m]	*n.*	棕榈树
sesame ['sesəmɪ]	*n.*	芝麻
safflower ['sæflauə]	*n.*	红花；［染料］红花染料
hempseed ['hempˌsi:d]	*n.*	大麻籽

perilla [pə'rɪlə]	*n.*	紫苏属
linseed ['lɪnˌsi:d]	*n.*	亚麻籽
linolenic [lɪnə'lenɪk] acid	*n.*	亚麻酸
melting points	*n.*	熔点
trisaturated glycerides	*n.*	三饱和酸甘油酯
satiety [sə'taɪətɪ]	*n.*	满足；饱足；过多
cholesterol [kə'lestərɒl]	*n.*	胆固醇
permeability [ˌpɜ:miə'bɪləti]	*n.*	渗透性
membrane ['membreɪn]	*n.*	细胞膜；薄膜；膜皮
thermal ['θɜ:ml]	*adj.*	热的；热量的
	n.	上升的热气流
insulation [ˌɪnsju'leɪʃən]	*n.*	绝缘；隔离，孤立
contributor [kən'trɪbjətə(r)]	*n.*	贡献者；捐助者；赠送者；投稿人；原因
medium ['mi:dɪəmˌ -djəm]	*n.*	方法；媒体；媒介；中间物
	adj.	中等的；适中的
shortening ['ʃɔ:tnɪŋ]	*n.*	起酥油
lubrication [ˌlu:brɪ'keɪʃən]	*n.*	润滑；润滑作用
emulsify [ɪ'mʌlsɪfaɪ]	*vt.*	使…乳化
	vi.	乳化
ingredient [ɪn'gri:dɪənt]	*n.*	成分；原料；配料；因素
mayonnaise [ˌmeɪə'neɪz]	*n.*	蛋黄酱
dressing ['dresɪŋ]	*n.*	穿衣；加工；调味品；装饰；打扮；填料；肥料
viscous ['vɪskəs]	*adj.*	黏性的；黏的
pourable ['pɔ:rəbl]	*adj.*	流动通畅的；可浇注的
confection [kən'fekʃən]	*n.*	糖果，蜜饯；调制；糖膏（剂）；精制工艺品
lecithin ['lesɪθɪn]	*n.*	卵磷脂；蛋黄素
emulsifiers [ɪ'mʌlsɪfaɪəz]	*n.*	乳化剂，黏合剂
staling ['steɪlɪŋ]	*n.*	老化；停滞
inhibitor [ɪn'hɪbɪtə(r)]	*n.*	抑制剂；抗化剂；抑制者
legume ['legju:m]	*n.*	豆科植物，豆类蔬菜
bleach [bli:tʃ]	*vt.*	使漂白；使变白
deodorize [di: 'əʊdəraɪz]	*vt.*	脱去…的臭味；防臭

hydrogenated [haɪ'drɒdʒəneɪtɪd]	*adj.*	氢化的；加氢的
margarine [ˌmɑːdʒə'riːn]	*n.*	人造黄油；人造奶油
blend [blend]	*vi.*	混合；协调
	n.	混合物
flavoring ['fleɪvərɪŋ]	*n.*	调味品，调味料
	v.	给…调味（flavor 的现在分词）
superiority [suːˌpɪərɪ'ɒrətɪ]	*n.*	优越，优势；优越性
atherosclerotic [ˌæθərəʊsklə'rɒtɪk]	*adj.*	动脉粥样硬化的
	n.	动脉粥样硬化患者
plasticity [plæ'stɪsətɪ]	*n.*	塑性，可塑性；柔软性
churn [tʃəːn]	*v.*	剧烈搅动；（使）猛烈翻腾；用搅乳器搅
	n.	（制作黄油的）搅乳器；（旧时）盛奶大罐
casein ['keɪsɪɪn]	*n.*	酪蛋白；干酪素
lactose ['læktəʊs]	*n.*	乳糖
calcium['kælsɪəm]	*n.*	钙
lactone ['læktəʊn]	*n.*	内酯
extender [ɪks'tendə]	*n.*	增量剂；补充剂；增充剂；延长器

Notes

1）Lipid：脂质、油脂，是油、脂肪、类脂的总称。食物中的油脂主要是油、脂肪，一般把常温下是液体的称作油，而把常温下是固体的称作脂肪。脂质是人体需要的重要营养素之一，它与蛋白质、碳水化合物是产能的三大营养素，在供给人体能量方面起着重要作用。

2）Phospholipid：磷脂，也称磷脂类、磷脂质，是指含有磷酸的脂类，属于复合脂。磷脂是组成生物膜的主要成分，分为甘油磷脂与鞘磷脂两大类，分别由甘油和鞘氨醇构成。磷脂常与蛋白质、糖脂、胆固醇等其他分子共同构成脂双分子层，即细胞膜的结构。

3）Triglyceride：甘油三酯，是三分子长链脂肪酸和甘油形成的脂肪分子。甘油三酯是人体内含量最多的脂类，大部分组织均可以利用甘油三酯分解产物供给能量，同时肝脏、脂肪等组织还可以进行甘油三酯的合成，在脂肪组织中贮存。

4）Cholesterol：胆固醇，又称胆甾醇。一种环戊烷多氢菲的衍生物。胆固醇是动物组织细胞所不可缺少的重要物质，它不仅参与形成细胞膜，而且是合成胆汁酸、维生素 D 及甾体激素的原料。

5）Margarine：人造黄油、人造奶油，用植物油加部分动物油、水、调味料经调配加工而成的可塑性的油脂品，用于代替从牛奶取得的天然奶油。

6）Lecithin：卵磷脂、蛋黄素，存在于每个细胞之中，更多的是集中在脑及神经系统、血液循环系统、免疫系统以及肝、心、肾等重要器官。卵磷脂可使大脑神经及时得到营养补充，有利于消除疲劳。卵磷脂具有乳化、分解油脂的作用，可增进血液循环，改善血清脂质，清除过氧化物。

Lesson 4 Vitamins and Minerals

Vitamins and minerals do not provide calories to the body. A few minerals serve as parts of body structures, but all vitamins and minerals act as regulators in all body processes. Vitamins are organic compounds, each with a unique chemical composition. Minerals are inorganic elements and are depicted in the periodic table. Unlike vitamins, minerals cannot be destroyed by heat, light, or oxygen.

Vitamins

Introduction

Vitamins are defined as "essential, low-molecular-weight, organic nutrients that are required in trace amounts in the diet", because they cannot be synthesized generally in the organism or the synthesis of the body cannot meet the needs. Though they are neither for body structure nor for energy supply, vitamins have important roles in maintaining physiological functions of promoting growth and regulating metabolism.

The term vitamine was introduced by Casimer Funk in 1911 to describe essential accessory chemical factors for life processes. According to Funk's belief, the antiberiberi factor and other factors were amines, but later studies did not substantiate this belief and the terminal "e" was eventually removed, leaving the word "vitamin".

It was found in the eighteenth century that small quantities of citrus juice would prevent the development of scurvy in seafaring people during long voyages. The vitamin necessary to prevent scurvy is ascorbic acid. Late in the nineteenth century, it was shown that beriberi resulted from the eating of polished rice. The vitamin to prevent beriberi is thiamin, now known to be present in the bran removed from the rice during the milling process.

Early in the twentieth century, it was discovered that for health and normal growth the usual nutrients, proteins, carbohydrates, and fats together with water and some mineral salts will not suffice. Other factors are necessary also. About the same time, these factors were called "vitamines" because of their supposed amino nature. That name has since been shortened to "vitamin". These compounds are of different compositions. Because of these differences in chemical composition, the following classification is used.

Anatomy of the Vitamins

Although many people think of vitamins as energy boosters, in truth, vitamins do not supply the body with energy in the form of calories—a fact that distinguishes them from fat, carbohydrate, and protein. However, many vitamins regulate the chemical reactions that allow us to obtain energy from those nutrients. Vitamins differ from fat, protein, and carbohydrate in other important ways. For one, the amounts of vitamins a body needs daily—a mere microgram or two in some cases—are infinitesimal compared to the grams of fat, carbohydrate, and protein required each day. Another difference is structural: vitamins are individual units rather than long chains of smaller units.

Like fat, carbohydrate and protein, however, vitamins are organic (carbon-containing) compounds essential for normal functioning, growth, and maintenance of the body. The functions of vitamins are often interrelated, so a deficiency of just one can cause profound health problems.

The vitamins are an untidy collection of complex organic nutrients which occur in the biological materials we consume as food (and in others we do not). In terms of chemical structure, they have nothing in common with one another, and their biological functions similarly offer little help with their definition or classification.

Category

Vitamins are divided into two groups: fat-soluble (lipid soluble) and water-soluble. Fat-soluble vitamins are comprised of vitamin A, D, E and K, which can be absorbed efficiently when a normal fat absorption exist. Fat-soluble vitamins are transported in lipoproteins or associated with special protein carriers in the blood. Like other fats, the fat-soluble vitamins can be stored in fatty tissues, excessive intake of vitamin A and D can accumulate to the toxic concentrations. The roles of the fat-soluble vitamins are ranged from each other, and they play considerable roles in a variety of critical biological processes.

Scientists categorize vitamins based on their solubility. Vitamins A, D, E and K are lipid-like molecules that are soluble in fat. The B vitamins and vitamin C, on the other hand, are soluble in water. This difference in solubility affects the way, which the body absorbs, transports, and stores vitamins.

Water-soluble vitamins are absorbed in the intestinal cell and delivered directly to the bloodstream. The kidneys filter out excesses of most water-soluble vitamins and excrete them in urine. Fat-soluble vitamins are absorbed with fat. Micelles carry fat-soluble vitamins and dietary fat to the brush border for absorption, then chylomicrons containing fat-soluble vitamins are transported via the lymph to the bloodstream and eventually to the liver.

Requirements

Vitamins are considered as essential biochemical substances that certain organisms, including man, are unable to synthesize at concentration levels required for growth and maintenance of normal health. Many substances labeled vitamins are vitamins in the true sense of the word only for certain organisms. Vitamin C (ascorbic acid) is a required exogenous nutrient for man because the body lacks the enzyme known as L-gulonoxidase which converts L-gulonolactone to L-ascorbic acid. Since most other higher plants and animals can manufacture ascorbic acid in required amounts, ascorbic acid cannot be considered a vitamin for these organisms.

For man and other animals, dietary sources provide requisite concentrations of most vitamins, while intestinal-tract flora complement the dietary availability of many B vitamins. Some vitamins such as niacin can be synthesized within the animal body through conversion of tryptophan to nicotinic acid, but the maintenance of vitamin requirements in this fashion is usually of little significance for other types of vitamins.

The dietary requirements for various vitamins are variable since humans and other animal subjects may display different requirements depending on their health, stage of growth, and living habits. A complete deficiency of a particular vitamin (avitaminosis) may be indicated as a gross or subclinical deficiency. From diagnostic perspectives, gross vitamin deficiencies such as advanced scurvy caused by lack of vitamin C, can be easily determined; but a marginal, subclinical deficiency for the same vitamin may be indicated only as a slow healing wound or ulcer in a subject who otherwise displays good health. Marginal deficiencies in other vitamins may become evident as headache, loss of appetite, insomnia, nervousness, and/or a host of many seemingly unrelated problems—all of which indicate no significant cause-effect relationship between the problem and the symptomatic effect. In spite of concerning gross vitamin deficiencies, the clinical and biochemical record is still disjointed and incomplete regarding the effects and diagnosis of these subclinical situations.

Storage and Toxicity

Fat-soluble vitamins accumulate in the liver and adipose tissues where they can be drawn upon in times of need. Once these vitamin stores are established, you can go for days, weeks, or even months without consuming more and suffer no ill effects. On the other hand, excessive intake of the fat-soluble vitamins A or D can exceed the body's storage capacity, with toxic effects.

Vitamin toxicity is rarely linked to high vitamin intakes from food or to the use of supplements that contain 100%-150% of the recommended amounts. People who take megadoses of one or more vitamins run a high risk of toxicity.

Provitamins

Certain vitamins in foods are in inactive forms that the body cannot use directly. These substances are known as provitamins, or vitamin precursors. Once a provitamin is ingested, the body converts it to the active vitamin form. One familiar provitamin in many fruits and vegetable is beta-carotene. Once beta-carotene is absorbed, the body converts it to an active form of vitamin A. In fact, beta-carotene is a major source of vitamin A in the diet. When experts calculate vitamin requirements or monitor consumption, they must take provitamins into account.

Minerals

Introduction

To form living protoplasm, the animal and vegetable organism needs minerals as well as carbohydrates, proteins, fats, vitamins and water. Like the purely organic materials, minerals help to create the correct physical and chemical conditions for the functioning of cells and tissues; the regulation of the osmotic pressure and the degree of water absorption by proteins. They control the development of the electric potential at the interface, activation or inhibition of enzyme systems and functioning of the buffer systems.

Minerals are also responsible for the slightly alkaline reaction of the blood and tissue fluids; they take part in growth and blood formation and are involved in many ways in the functioning and synthesis of hormones, vitamins and enzymes. They are especially important as building blocks for bone, teeth and certain tissues. They are subject to constant change and loss similar to the rest of body's building blocks, and the body has to make up these losses.

Unlike the nutrient molecules you have studied so far, minerals are inorganic elemental atoms or ions. Unlike carbohydrate, protein, and fat, minerals are not changed during digestion or when the body uses them. Unlike many vitamins, minerals are not destroyed by heat, light, or alkalinity. Calcium remains calcium, be it in seashells, milk, or bones. Iron remains iron, whether it is part of a cast-iron skillet or carried in the bloodstream as part of hemoglobin. This is true for all minerals.

Although we often associate minerals with animal foods (e.g., calcium in dairy products, iron in meats), minerals are found throughout the groups that make up the food guide pyramid (FGP). Generally animal foods are more reliable sources of minerals than plants are, because animal tissues contain minerals in the proportions that animal tissues need. Drinking water can sometimes be a significant source of several minerals such as sodium, magnesium, and fluoride.

Category

Minerals are essential inorganic elements. Those that we need and store in larger amounts are called major minerals or macrominerals, and those we need in very small quantities are the trace minerals. Major minerals include calcium, phosphorus, magnesium, sodium, potassium, chloride and sulfur. Trace minerals include iron, manganese, copper, iodine, zinc, cobalt, fluoride and selenium.

Functions

Minerals play an essential role in the body. Some minerals, such as magnesium, participate in the catalytic activity of enzymes. Others serve a structural function, for example, calcium and phosphorus are among the minerals that make our bones hard. Minerals are categorized as major or trace minerals, based on the amount needed in the diet and the amount of the mineral in the body. The body requires more than 100 mg/d of each major mineral, while the dietary need for each trace mineral is less than 100 mg/d.

Calcium is especially important as it only occurs in a few foods in sufficient amounts. Calcium deficiency is the most frequent mineral deficiency in nutrition. Calcium ions and phosphoric acid in combination with vitamin D must be available in the right proportions for the formation of bone. Not all the calcium in food is adsorbed, and one example is the calcium phosphate in cereals. The type of bonding is important. Absorption is also dependent on the protein content of a diet. Easily soluble calcium salts, such as chloride, lactate and gluconate are absorbed as well as the less soluble calcium sulphate, calcium phosphate and calcium carbonate. A normal diet must therefore contain an adequate supply of calcium, which is why many countries enrich certain foods (bread, flour) with calcium in the form of calcium carbonate.

Sodium and chlorine ions are present in food mainly in the form of salt. They make food taste better; regulate the osmotic pressure of the body fluids, form acid in the mucous membranes of the stomach and keep digestive processes normal. Potassium has a similar function.

The metabolic requirements for phosphoric acid (with calcium and vitamin D), is not only important for the formation of bone, but the phosphates and polyphosphates are essential in the intermediate metabolism and energy processes of all living cells.

The so-called trace elements, which are present in the body in very small amounts, have important functions and therefore the body must receive adequate supplies. Serious symptoms result from a deficiency of these elements. Iron and copper are important elements in some enzyme systems. Iodine is needed for the formation of the hormone thyroxin in the thyroid gland. Manganese is the co-enzyme of arginase and alkaline phosphatase. Zinc is a component of the pancreatic hormone, insulin and of certain enzymes of yeast such as carbonic anhydrase, carboxy polypeptidase and alcohol dehydrase. Vitamin B_{12} contains cobalt needed for red blood cell formation. Molybdenum is a constituent of xanthinoxidase and also necessary for the fixation of nitrogen in the air by soil bacteria and is therefore essential for life. *Aspergillus niger* needs gallium to grow. A lack of boron leads, in the higher plants, to severe disturbance of growth, for instance, to rotting of the core of sugar beet. Selenium which is toxic in large amounts is now considered to be of physiological importance as a trace element. Zinc is probably essential in trace amounts.

Aluminium, nickel, chromium, titanium and uranium are still considered to be of doubtful importance. It is not known what part fluorine plays, but presumably it helps protect the teeth. Fluorine inhibits caries, but caries is not a deficiency disease of fluorine, but is due to bacterial attack. Lead, mercury, arsenic, antimony, thallium and the rare earths are toxic elements.

New Words

vitamin ['vɪtəmɪn]	*n.*	维生素
mineral ['mɪnərəl]	*n.*	矿物；矿物质；矿泉水；无机物；苏打水
	adj.	矿物的；矿质的
organic [ɔː'gænɪk]	*adj.*	有机的；组织的；器官的；根本的
	n.	有机物
polished rice		精米；白米
nutrient ['njuːtrɪənt]	*n.*	营养素；营养物；滋养物
	adj.	营养的；滋养的
ascorbic acid		抗坏血酸；维生素 C
toxicity [tɒk'sɪsətɪ]	*n.*	毒性；毒力
provitamin [prəʊ'vaɪtəmɪn]	*n.*	维生素原；前维生素

Notes

1) Vitamins are defined as "essential, low-molecular-weight, organic nutrients that are required in trace amounts in the diet", because they cannot be synthesized generally in the organism or the synthesis of the body cannot meet the needs.
   参考译文：维生素是一类人体必需的小分子的有机营养素。因为有机体不能合成或者所合成的维生素满足不了需求，所以在膳食中需要少量的维生素。
2) Like the purely organic materials, minerals help to create the correct physical and chemical conditions for the functioning of cells and tissues; the regulation of the osmotic pressure and the degree of water absorption by proteins.
   参考译文：与那些纯有机物类似，矿物质有助于创造发挥细胞和组织功能所需的恰当的理化条件，调节渗透压和蛋白质所吸收的水分。

Lesson 5 Food Nutrition and Malnutrition

Food Nutrition

Although we give food meaning through our culture and experience and make dietary decisions based on a whole host of factors, ultimately the reason for eating is to obtain nourishment—nutrition. Nutrition is the science that interprets the interaction of nutrients and other substances in food (e.g., phytonutrients, anthocyanins, tannins) in relation to maintenance, growth, reproduction, health and disease of an organism. It includes food intake, absorption, assimilation, biosynthesis, catabolism and excretion.

Nutrition is the provision, to cells and organisms, of the materials necessary (in the form of food) to support life. Many common health problems can be prevented or alleviated with good nutrition. Just like our body, food is a mixture of chemicals, some of which are essential for normal body function. These essential chemicals are called nutrients.

Nutrients

Nutrient is any element or compound necessary for contributing to an organism's metabolism, growth, or other functioning. We need nutrients for normal growth and development, for maintaining cells and tissues, for fuel to do physical and metabolic work, and for regulating the hundreds of thousands of body processes that go on inside our every second of every day. Further, food must provide these nutrients which the body either cannot make or cannot make adequate amounts of the essential nutrients.

There are seven major classes of nutrients: carbohydrates, lipids (fats and oils), proteins, fibers, minerals, vitamins, and water. These nutrient classes can be generally grouped into the categories of macronutrients, and micronutrients. The macronutrients are carbohydrates, fats, fibers, proteins and water. The other nutrient classes are micronutrients. The macronutrients (excluding fiber and water) provide energy, which is measured in kilocalories, often called "calories". Carbohydrates and proteins provide four calories of energy per gram, while fats provide nine calories per gram. Vitamins, minerals, fiber, and water do not provide energy, but are necessary for other reasons. Other micronutrients not categorized above include antioxidants, essential fatty acids, and phytochemicals. Most foods contain a mix of some or all of the nutrient classes.

Nutritional Assessment

Nutritional assessment (NA) is the first step in the treatment of malnutrition. Specific data are obtained to create a metabolic and nutritional profile of the patient. The goals of NA are identification of patients who have, or are at risk of developing malnutrition, to quantify a patient's degree of malnutrition, and to monitor the adequacy of nutrition therapy.

Environmental Nutrition

Research in the field of nutrition has greatly contributed in finding out the essential facts about how environmental depletion can lead to crucial nutrition-related health problems like contamination, spread of contagious diseases, malnutrition, etc. Moreover, environmental contamination due to discharge of agricultural as well as industrial chemicals like organocholrines, heavy metal, and radionucleotides may adversely affect the human and the ecosystem as a whole. As far as safety of the human health is concerned, then these environmental contaminants can reduce people's nutritional status and health. This could directly or indirectly cause drastic changes in their diet habits. Hence, food-based remedial as well as preventive strategies are essential to address global issues like hunger and malnutrition and to enable the susceptible people to adapt themselves to all these environmental as well as socio-economic alterations.

Malnutrition: A National Problem

Malnutrition refers to insufficient, excessive, or imbalanced consumption of nutrients by an organism. In developed countries, the diseases of malnutrition are most often associated with nutritional imbalances or excessive consumption. In developing countries, malnutrition is more likely to be caused by poor access to a range of nutritious foods or inadequate knowledge.

Although there are more organisms in the world who are malnourished due to insufficient consumption, increasingly more organisms suffer from excessive over-nutrition; a problem caused by an over-abundance of sustenance coupled with the instinctual desire (by animals in particular) to consume all that it can.

(1) Insufficient: In general, under-consumption refers to the long-term consumption of insufficient sustenance in relation to the energy that an organism expends or expels, leading to poor health.

(2) Excessive: In general, over-consumption refers to the long-term consumption of excess sustenance in relation to the energy that an organism expends or expels, leading to poor health and, in animals, obesity. It can cause excessive hair loss, brittle nails, and irregular premenstrual cycles for females.

(3) Unbalanced: When too much of one or more nutrients are present in the diet to the exclusion of the proper amount of other nutrients, the diet is said to be unbalanced.

It would be erroneous to conclude that only people who live at or below the poverty level suffer from malnutrition, and hence are susceptible to underdevelopment physically and mentally. According to the food consumption survey conducted by USDA's Agricultural Research Service in 1965, over one-third of the households with incomes of $10,000 or more did not have diets that met all recommended levels of all the nutrients to provide a good diet, and 9% of the families in this income bracket had diets rated as "poor" actually. As the family income declined, so did the diet rating. At an income level of $3,000 or less, 36% of the households had diets rated as poor. Food likes and dislikes, food fads, ethnic backgrounds, habits, and incomes all influence the dietary patterns of rich and poor alike. It is therefore evident that to supply merely an abundance of food to combat malnutrition would be only a partial attack upon a complex problem. It has long been known that if a food supplement is to be successful in nourishing a malnourished population, it must be acceptable to the people for whom it is intended. Changing food fads and habits even in malnourished populations is extremely difficult. Therefore, nutrition education is of the utmost importance to any nutrition program whether in the United States (US) or in other countries.

The National School Lunch Program—a Remedy

The National School Lunch Program offers several approaches to solving the malnourishment problem:

(1) The nutritive content of the meal must meet at least a third of the child's nutritional requirements for the day, containing all the elements essential to a balanced meal.

(2) Through federal, state and local support, the price of the meal is within the ability of most of the children to pay.

(3) By federal regulation, children who are unable to pay the full price of the meal must be provided a lunch free of charge or at a reduced price.

(4) The menu pattern is devised to give extensive latitude to the local schools in planning the meals from day to day; yet the pattern will provide the full nutritional requirements when adhered to with a wide variety of foods to choose from.

(5) Even though local food habits and patterns are observed in menu planning, the program provides an excellent opportunity for introducing foods which the children are not accustomed to eating at home and which will broaden their range of selection to help insure an adequate and balanced diet.

(6) The day-to-day participation in the program develops good food habits which will carry on through adulthood and into the community.

(7) Properly coordinated with classroom work, the lunchroom can be a laboratory for actual experience in the principles of nutrition, sanitation, safety, personal hygiene, food service management, courtesies and social graces, budgeting, accounting, food storage and handling, food preservation, delivery systems, and many other subjects of importance to society.

New Words

nutrition [njʊ'trɪʃ (n)]	*n.*	营养，营养学；营养品；养分
malnutrition [ˌmælnjʊ'trɪʃ(ə)n]	*n.*	营养失调，营养不良
nourishment ['nʌrɪʃm(ə)nt]	*n.*	食物；营养品；滋养品；养料
macronutrient [ˌmækrəʊ'njuːtrɪənt]	*n.*	大量营养素；常量营养元素；宏量营养素

micronutrient [maɪkrə(ʊ)'nju:trɪənt]	*n.*	微量营养素
phytochemicals [faɪtəʊ'kemɪklz]	*n.*	植物化学物；植物素（复数）
	adj.	植物化学的
antioxidant [ˌæntɪ'ɒksɪdənt]	*n.*	抗氧化剂；防老化剂
sanitation [ˌsænɪ'teɪʃn]	*n.*	卫生，公共卫生；环境卫生；卫生设备
hygiene['haɪdʒi:n]	*n.*	卫生；卫生学，保健学；环保

Notes

1) We need nutrients for normal growth and development, for maintaining cells and tissues, for fuel to do physical and metabolic work, and for regulating the hundreds of thousands of body processes that go on inside our every second of every day.

参考译文：我们需要营养素来进行正常生长发育，维持细胞和组织，为新陈代谢提供能量，调节那些时刻进行着的成千上万的身体进程。

2) In developed countries, the diseases of malnutrition are most often associated with nutritional imbalances or excessive consumption. In developing countries, malnutrition is more likely to be caused by poor access to a range of nutritious foods or inadequate knowledge.

参考译文：在发达国家，营养不良大多与营养不平衡或者过度消耗相联系。在发展中国家，营养不良可能更多是由摄入有营养的食物不足或者营养常识不正确引起的。

Reading Material 1 Dietary Fiber

Dietary fiber is that part of plant material in the diet which is resistant to enzymatic digestion, includes cellulose (*n.* 纤维素), noncellulosic (*adj.* 非纤维素的) polysaccharides (*n.* 多糖) such as hemicellulose (*n.* 半纤维素), pectic (*n.* 果胶) substances, gums, mucilages (*n.* 黏液) and a non-carbohydrate component lignin (*n.* 木质素). The diets rich in fiber such as cereals, nuts, fruits and vegetables have a positive effect on health, since their consumption has been related to decrease incidence of several diseases. Dietary fiber can be used in various functional foods like drinks, bakery and meat products. Influence of different processing treatments, like extrusion-cooking (*n.* 挤压烹饪), canning (*n.* 灌装), grinding (*n.* 研磨), boiling, frying) alters (*v.* 改变) the physico chemical properties of dietary fiber and improves their functionality. Dietary fiber can be determined by different methods, mainly by enzymic gravimetric (*adj.* 酶重量法) and enzymic-chemical methods. Dietary fibers are a complex group of carbohydrates and lignin that are not hydrolyzed (*v.* 水解) by human enzymes, and therefore, are not digested or absorbed in the human body. Dietary fiber is intact in plants and is composed of a complex polymer (*n.* 聚合体) of phenylpropanoid (*n.* 苯丙烷) subunits (*n.* 亚基). Soluble fiber is the edible part of plant that is resistant to digestion but could be partially or totally fermented by colonic bacteria to short-chain fatty acids in the large intestine. Meanwhile, insoluble fiber passes through the digestive tract intact.

Issues Arising from the Definition of Dietary Fiber

American Association of Cereal Chemists (AACC) (美国谷物化学师学会) in 2000 defined dietary fiber as the edible parts of plant or analogous (*adj.* 与…相类似的) carbohydrates that are resistant to digestion and absorption in the human small intestine (*n.* 肠) with complete or partial fermentation in the large intestine. Dietary fiber includes polysaccharides, oligosaccharides (*n.* 低聚糖), lignin and associated plant substances. During the year 2001, Australia New Zealand Food Authority (ANZFA) (澳大利亚新西兰食品标准局) defined dietary fiber as that fraction of the edible parts of plants or their extracts, or analogous carbohydrates, that are resistant to digestion and absorption in the human small intestine, usually with complete or partial fermentation in the large intestine. The term includes polysaccharides, oligosaccharides and lignin. The panel on the definition of dietary fiber constituted by National Academy of Science (NAS) during the year 2002 defined that the dietary fiber complex includes dietary fiber and functional fibers. Dietary fiber consists of non-digestible carbohydrates and lignin that are intrinsic and intact in plants, and functional fibers consists of isolated, non-digestible carbohydrates which have beneficial physiological effects in humans and total fiber as the sum of dietary fiber and functional fiber.

Categorization of Fiber in Foods

The most widely accepted approach to classify fiber in foods is to differentiate the different forms based on (1) their solubility (*n.* 溶解度) in a buffer at a defined pH, and/or (2) their fermentability in an *in vitro* (*n.* 体外) system mimicking (*v.* 模仿) human alimentary (*n.* 消化道)enzymes. Since most fiber types are at least partially fermented, it is also now commonly accepted to classify fiber as partially or poorly fermented, and well-fermented. Generally (not universally), well-fermented fibers are soluble in water, while partially or poorly fermented fibers are insoluble. There are other classified systems such as those based on the role of fiber in the plant, the type of polysaccharide, the degree of simulated gastrointestinal (*n.* 胃肠道) fermentability, the site of digestion, and others based on products of digestion and physiological classification.

There are complications (*n.* 并发症, 多样的) with each classification system, as dietary fiber fractions consist of a wide range of different compounds, each with a unique chemical structure and physical properties. More work is needed in this area in order to elucidate if there is a more effective classification system than those that are currently accepted in the scientific literature and in particular, if further differentiation by classification would better describe or group together those fibers with similar physiological properties.

Health Effects of Dietary Fiber

Epidemiologically (*adv.* 流行病学上地), the study of dietary fiber has come from the study of whole foods (not individual fibers), which form the basis of a diet. A recent meta-analysis found that high consumption of whole grains or cereal fiber was significantly associated with reduced risk of all-cause mortality (*n.* 死亡率) and death from cardiovascular disease (*n.* 心血管疾病), cancer, diabetes (*n.* 糖尿病), respiratory (*adj.* 呼吸的) disease, infections (*n.* 传染病), and other causes, suggesting that the protective effects of whole grains may be due, at least in the main part, to its cereal fiber component. As whole grain cereals are a rich source of fiber and other bioactive compounds, the precise physiological effects exerted by whole grain cereals are still being elucidated, however studies have suggested that it is dietary fiber that largely determines the quality of cereal foods. Fiber can also help to reduce the energy density of foods owing to its bulking (*n.* 膨胀) effects and promoting satiety (*n.* 饱腹). Effects of other fibers in other foods, namely fruit and vegetables, and not surprisingly chemical adaptations (*n.* 适应性) of naturally occurring fibers known as synthetic fibers, are also relevant although the effects are less well elucidated.

The following physiological effects have been attributed to dietary fiber: Increased fecal (*n.* 粪便) bulk (*n.* 松散), reduced total serum (*n.* 血清) cholesterol (*n.* 胆固醇) levels, attenuation of postprandial (*adj.* 餐后的) glycaemia (*n.* 血糖), reduced blood pressure, decreased transit time, increased colonic fermentation/short chain fatty acid production, positive modulation of colonic microflora, weight loss/reduction in adiposity, increased satiety, beneficial effect on mineral absorption, a protective role in the prevention of colon cancer.

Dietary fiber has several protective effects against chronic diseases, including cardiovascular disease, diabetes, metabolic syndrome (*n.* 综合征), inflammatory (*n.* 炎症) bowel syndrome, obesity, and colorectal cancer in the age-adjusted analysis. For example, insoluble fiber binds to and adsorb carcinogens (*n.* 致癌物质), mutagens (*n.* 诱变剂), and toxins (*n.* 毒素), and therefore, prevents their harmful effects to the body by preventing the toxins absorption and targeting them for elimination. Other fiber properties include delayed colonic transit time, prolonged post-meal satiety and satiation, and induction of cholecystokinin (*n.* 胆囊收缩素) satiety hormone. The Academy of Nutrition and Dietetics position on fiber intake is to increase consumption of whole grains, fruits and vegetables, nuts and legumes (*n.* 豆类), and that dietary fiber is associated with risk reduction of type Ⅱ diabetes, cardiovascular disease, and select cancer types.

Recommended Dietary Fiber Intakes

The recommended dietary reference intake (DRI) daily allowance for men aged 19-50 years is 38 g/d and women 25 g/d, for men ages > 51 is 31 g/d and women ages > 51 is 21 g/d. The recommendation for children ages 1-3 is 19 g/d and ages 4-8 is 25 g/d. For boys, ages 9-13, the DRI recommendations are 31 g/d, and 38 g/d for ages 14-18. For girls ages 9-18, the DRI recommendations are 26 g/d. Although dietary fibers have been shown have several beneficial health effects, the average daily intake for most Americans is 15 g/d, which is much lower than the recommended amount. There is no upper tolerable level for fiber intake, but the tolerance varies by individual, and the most common side effects from overconsumption are bloating (*n.* 腹胀) and abdominal (*n.* 腹部) discomfort.

Concluding Remarks

To conclude, much evidence supports an important role for dietary fiber intake as a contributor to overall metabolic health, through key pathways that include insulin (*n.* 胰岛素) sensitivity. Furthermore, there are clear associations between dietary fiber intake and multiple pathologies that include cardiovascular disease, colonic health. Dietary fiber intake also correlates with mortality. The gut microflora functions as an important mediator of the beneficial effects of dietary fiber, including the regulation of appetite and metabolic processes and chronic inflammatory pathways. Many factors contribute towards the impoverishment (*n.* 贫瘠) of dietary fiber intake in the typical Western diet. Unfortunately, there is habituation of many of us to our modern-day environs, lifestyles, diets and eating-related behaviors. The problem is that what most of us consider normal is actually highly abnormal and about as far away from what our hominid (*n.* 原始人) hunter-gatherer ancestors experienced and enjoyed as it is possible to imagine. The fact is that over decades, a blink of an eye in hominid history, we have gradually migrated into our current environments, lifestyles and culture. Stepwise changes are required. By adopting some of the suggested strategies here, it is our belief that real change in dietary fiber intake can occur.

As food consumers, our choice of high-fiber foods in preference to fiber-impoverished ultra-processed foods likely has a major positive impact on our future health and wellbeing and will ultimately influence the strategic commercial plans of food companies, with likely future improvements in the fiber content of processed food production. In the capitalist culture of modern Westernized societies with 'consumer as king', we all need to vote with our mouths, and in the process, re-discover the joy of cooking with fresh and fiber-replete ingredients.

Notes

1) Dietary fiber is that part of plant material in the diet which is resistant to enzymatic digestion, includes cellulose, noncellulosic polysaccharides such as hemicellulose, pectic substances, gums, mucilages and a non-carbohydrate component lignin.

参考译文：膳食纤维是饮食中对酶具有抗消化性的植物性组分，包括纤维素、非纤维素多糖（如半纤维素、果胶物质、树胶、黏液）和非碳水化合物组分木质素。

句中第一个“which”引导定语从句，用“resistant to enzymatic digestion”修饰“dietary fiber”。

2) Effects of other fibers in other foods, namely fruit and vegetables, and not surprisingly chemical adaptations of naturally occurring fibers known as synthetic fibers, are also relevant although the effects are less well elucidated.

参考译文：其他食物（即水果和蔬菜）中的纤维素，在人体中并不表现出奇的化学适应性，天然存在的纤维和人工合成的纤维也具有一定的功能相关性，尽管这些功能的研究尚不明确。

句中“namely fruit and vegetables”为插入语，用于修饰前文“Effects of other fibers in other foods”。

3) During the year 2001, Australia New Zealand Food Authority (ANZFA) defined dietary fiber as that fraction of the edible parts of plants or their extracts, or analogous carbohydrates, that are resistant to digestion and absorption in the human small intestine, usually with complete or partial fermentation in the large intestine.

参考译文：在 2001 年，澳大利亚新西兰食品管理局将膳食纤维定义为植物或其提取物或类似碳水化合物的可食用部分，其对人类小肠中的消化和吸收具有抗消化性，而在大肠可部分或完全发酵的植物或其提取物或类似碳水化合物的可食用部分。

句中“defined…as…：”将…定义为…，句中“that”引导定语从句，引导词之后部分为“dietary fiber”的定语。

4) There are complications with each classification system, as dietary fiber fractions consist of a wide range of different compounds, each with a unique chemical structure and physical properties.

参考译文：每种分类系统都存在复杂性，因为膳食纤维组分由多种不同的化合物组成，每种化合物具有独特的化学结构和一系列潜在的生理效应。

句中使用“there be”句型，“as”引导原因状语从句。

5) More work is needed in this area in order to elucidate if there is a more effective classification system than those that are currently accepted in the scientific literature and in particular, if further differentiation by classification would better describe or group together those fibers with similar physiological properties.

参考译文：在这一领域需要做更多的工作，以阐明是否存在比科学文献中目前接受的更有效的分类系统，特别是未来的分类系统是否能更好地描述和区分生理性质相近的膳食纤维。

句中“if”引导的是虚拟语气。

6) As whole grain cereals are a rich source of fiber and other bioactive compounds, the precise physiological effects exerted by whole grain cereals are still being elucidated, however studies have suggested that it is dietary fiber that largely determines the quality of cereal foods.

参考译文：由于全谷类谷物是纤维和其他生物活性化合物的丰富来源，全谷类谷物所产生的精确生理效应尚需进一步的阐明，但现有的研究已经表明，膳食纤维在很大程度上决定了谷类食品的质量。

句中“As”引导的是原因状语从句。

Exercise

1. Translate the following sentences into English.

1）同时，不溶性纤维完整地通过消化道。

2）可溶性纤维是植物的可食用部分，具有抗消化性，但可能被结肠细菌部分或完全发酵为大肠中的短链脂肪酸。

3）其他纤维特性包括结肠运输时间延迟、餐后饱腹感和饱腹感延长以及胆囊收缩素饱腹激素的诱导。

4）纤维摄入量没有最高的耐受水平，但耐受性因人而异，过度摄入最常见的副作用是腹胀和腹部不适。

5）在以“消费者为王”的现代西化社会资本主义文化中，我们都需要用嘴投票。在此过程中，重新发现了用新鲜和富含膳食纤维的食物烹饪的乐趣。

2. Translate the following sentences into Chinese.

1) There are other classification systems such as those based on the role of fiber in the plant, the type of polysaccharide, the degree of simulated gastrointestinal fermentability, the site of digestion, and others based on products of digestion and physiological classification.

2) The following physiological effects have been attributed to dietary fiber: Increased fecal bulk, reduced total serum cholesterol levels, attenuation of postprandial glycaemia, reduced blood pressure, decreased transit time, increased colonic fermentation/short chain fatty acid production, positive modulation of colonic microflora, weight loss/reduction in adiposity, increased satiety, beneficial effect on mineral absorption, a protective role in the prevention of colon cancer.

3) As food consumers, our choice of high-fiber foods in preference to fiber-impoverished ultra-processed foods likely has a major positive impact on our future health and wellbeing and will ultimately influence the strategic commercial plans of food companies, with likely future improvements in the fiber content of processed food production.

4) The fact is that over decades, a blink of an eye in hominid history, we have gradually migrated into our current environments, lifestyles and culture. Stepwise changes are required.

5) The Academy of Nutrition and Dietetics position on fiber intake is to increase consumption of whole grains, fruits and vegetables, nuts and legumes, and that dietary fiber is associated with risk reduction of type Ⅱ diabetes, cardiovascular disease, and select cancer types.

3. True or false.

1) The dietary fiber is not good for people's health. (　　)

2) The more dietary fiber we eat, the healthier we will be. (　　)

3) The dietary fiber can be used in people's stomach. (　　)

4) The old age people need more dietary fiber than young people need. ( )

5) The dietary fiber can help people defend cancer. ( )

6) The dietary fiber is defined complex to include dietary fiber consisting of non-digestible carbohydrates and lignin that are intrinsic and intact in plants, functional fibers consisting of isolated, non digestible carbohydrates which have beneficial physiological effects in humans and total fiber as the sum of dietary fiber and functional fiber in recent study. ( )

7) Because of the life change, people in foreign country eat less dietary fiber than their ancestors. ( )

8) Much evidence supports an important role for dietary fiber intake as a contributor to overall metabolic health, through key pathways but not include insulin sensitivity. ( )

Reading Material 2 Prebiotics and Postbiotics

Improving human health through modulation of microbial interactions during all phases of life is an evolving concept that is increasingly important for consumers, food manufacturers, health-care professionals and regulators. Microbiota-modulating dietary interventions include many fermented foods and fibre-rich dietary regimens, as well as probiotics, prebiotics and synbiotics, some of which are available as drugs and medical devices, as well as foods.

The past few decades have demonstrated unequivocally the importance of the human microbiota to both short-term and long-term human health. Early programming of the microbiota and immune system during pregnancy, delivery, breast-feeding and weaning is important and determines adult immune function, microbiome and overall health. We have also seen rapid growth in the number of products that claim to affect the functions and composition of the microbiota at different body sites to benefit human health.

Prebiotics and Probiotics

Consensus definitions of probiotics, prebiotics and synbiotics have been published previously. Probiotics are "live microorganisms that, when administered in adequate amounts, confer a health benefit on the host", whereas a prebiotic is a "substrate that is selectively utilized by host microorganisms conferring a health benefit". A synbiotic, initially conceived as a combination of both probiotics and prebiotics, has now been defined as "a mixture comprising live microorganisms and substrate(s) selectively utilized by host microorganisms that confers a health benefit on the host".

Prebiotics refer to substrates selectively utilized by beneficial microorganisms of the host, which are beneficial to health. Prebiotics must meet three criteria: resist digestion in the gastrointestinal tract (GIT); be fermentable by gut microbes; and promote the growth and/or activity of beneficial intestinal flora. Common prebiotics include inulin, fructo-oligosaccharides (低聚果糖), galacto-oligosaccharides (GOS)(低聚半乳糖), cello-oligosaccharides (纤维低聚糖), and some glucans. Prebiotics exert effects by improving intestinal flora thereby promoting epithelial barrier integrity or gut immunity. Prebiotics also serve as fermentation substrates for microorganisms, promoting the growth of specific bacteria, producing active substances, and improving intestinal microecology.

On the other hand, some studies suggest that prebiotics can be act as an active substance in direct contact with the surface recognition receptors of intestinal dendritic cells, there by exerting its effect independent of intestinal flora.

Unlike probiotics, prebiotics promote the growth of one or more beneficial bacteria endogenous to the host, thereby improving the gut microbiome. However, there are still some studies that have shown that excessive supplementation of prebiotics can lead to the overgrowth of probiotics, which can indirectly cause intestinal flora disturbance. Similar to probiotics, prebiotic intervention in food allergy should also be supplemented within a safe dose range based on the patient's health status.

Definition and Characteristics of Prebiotics

The term prebiotics is relatively new which is defined as "non-digestible food ingredients that beneficially affect the host by selectively stimulating the growth and/or activity of one or a limited number of bacteria in the colon for improving the host health". The definition of prebiotics has been reviewed by panel of experts from International Scientific Association for Probiotics and Prebiotics (ISAPP). The panel updated the definition in wider perspective as "a substrate that is selectively utilized by host microorganisms conferring a health benefit". This definition expands the concept of prebiotics to possibly include non-carbohydrate substances, applications to body sites other than the GIT and diverse categories other than food. These are generally short-chain carbohydrates which escape digestion but used as substrates for the growth of probiotics in upper GIT. Other compounds that are not classified as carbohydrates but are recommended to be prebiotics are cocoa-derived flavanols. *In vivo* and *in vitro* experiments demonstrate that flavanols can stimulate the growth of lactic acid bacteria .

Prebiotics broadly occur in several plants such as onion, asparagus (*n.* 芦笋), garlic, chicory (*n.* 菊苣), Jerusalem artichoke (*n.* 菊芋), oat and wheat which induce the existing metabolic activities in the colon by stimulating bacterial growth in the gut . It has been reported that there was a significant increase in fecal bifidobacterial population after consumption of fructo-oligosaccharides. Prebiotics are generally not hydrolyzed in the intestine, and they consist of low degrees of polymerization (2-20 units) in which monomers are generally glucose, galactose, fructose and/or xylose. Furthermore, they have caloric value due to non-digestion in the colon and have an energy contribution due to their involvement in fermentation. The most commonly studied prebiotics include fructo-oligosaccharides, isomalto-oligosaccharides (IMO)

(异麦芽寡糖) and xylo-oligosaccharides (XOS)(低聚木糖).

Postbiotics Definition

In 2019, ISAPP convened a panel of experts to review the definition and scope of postbiotics. The panel defined a postbiotic as a “preparation of inanimate microorganisms and/or their components that confers a health benefit on the host”. Effective postbiotics must contain inactivated microbial cells or cell components, with or without metabolites, that contribute to observed health benefits.

Probiotics are by definition alive and required to have an efficacious amount of viable bacteria at the time of administration to the host, but most probiotic preparations, especially at the end of shelf life, will also include potentially large numbers of dead and injured microorganisms. The potential influence of non-viable bacterial cells and their components on probiotic functionality has had little attention.

Characteristics and Effects of Postbiotics

Fermented foods might contain a substantial number of non-viable microbial cells, particularly after prolonged storage or after processing, such as pasteurization (e.g., soy sauce) or baking (e.g., sourdough bread). Food fermentation has a major influence on the physical properties and potential health effects of many foods, especially milk and plant-based foods. Many fermentations are mediated by lactic acid bacteria, which can produce a range of cellular structures and metabolites that have been associated with human health, including various cell surface components, lactic acid, short-chain fatty acids and bioactive peptides among other metabolites. These effector molecules of fermented food are thought to be similar to those product by probiotics, but this link has not been conclusively established.

Bacterial lysates of common bacterial respiratory pathogens have been used for decades to prevent paediatric respiratory diseases by postulated general immune-stimulating mechanisms that are not yet well understood. The possibility could be that nonviable microorganisms, their components and their end-products contributing to the health. We consider that a common understanding of the emerging concept of postbiotics, including a consensus definition, would benefit all stakeholders and facilitate developments of this field. Several aspects of postbiotics have been discussed, including processing factors important in their creation, proper characterization, mechanistic rationale on how they work to improve both intestinal and systemic health, safety and current regulatory frameworks. Key conclusions from this consensus panel are provided below.

(1) A postbiotic is defined as a “preparation of inanimate microorganisms and/or their components that confers a health benefit on the host”.

(2) Postbiotics are deliberately inactivated microbial cells with or without metabolites or

cell components that contribute to demonstrated health benefits.

(3) Purified microbial metabolites and vaccines are not postbiotics.

(4) A postbiotic does not have to be derived from a probiotic for the inactivated version to be accepted as a postbiotic.

(5) The beneficial effects of a postbiotic on health must be confirmed in the target host (species and subpopulation).

(6) The host can include humans, companion animals, livestock and other targets.

(7) The site of action for postbiotics is not limited to the gut. Postbiotics must be administered at a host surface, such as the oral cavity, gut, skin, urogenital tract (*n.* [解剖]泌尿生殖道) or nasopharynx (*n.* [解剖]鼻咽). Injections are outside the scope of postbiotics.

(8) Implicit in the definition of a postbiotic is the requirement that the postbiotic is safe for the intended use.

Overall, the safety and potential harms of postbiotic interventions remain poorly explored. Further studies are necessary to determine the effects and safety of different postbiotics.

Notes

1) Microbiota-modulating dietary interventions include many fermented foods and fibre-rich dietary regimens, as well as probiotics, prebiotics and synbiotics, some of which are available as drugs and medical devices, as well as foods.

参考译文：调节微生物菌群的膳食干预措施包括：采用大量发酵食品和富含纤维的饮食规则，以及可以从药物、医疗设备、食品中获得的益生菌、益生元和合生元。句中“as well as”表示并列关系，可翻译为：和…一样；不但…而且。

2) Fermented foods might contain a substantial number of non-viable microbial cells, particularly after prolonged storage or after processing, such as pasteurization (for example, soy sauce) or baking (for example, sourdough bread). Food fermentation has a major influence on the physical properties and potential health effects of many foods, especially milk and plant-based foods.

参考译文：发酵食品可能含有大量不可活的微生物细胞，特别是在长时间贮存或加工后，如巴氏杀菌（如酱油）或烘焙（如酵母面包）。食品发酵对许多食品的物理性质和潜在健康性质都有重大影响，尤其是牛奶和植物性食品。

句中“especially”其后可接名词、介词短语、从句等，表示陈述某一事实之后，列举一个具有代表性的例子，作进一步强调。

Exercise

1. True or false.

1) Only probiotics are helpful to intestinal flora regulation. ( )

2) Postbiotics are deliberately inactivated microbial cells with or without metabolites or cell components that contribute to demonstrated health benefits. ( )

2. Translate the following sentences into Chinese.

1) Prebiotics refer to substrates selectively utilized by beneficial microorganisms of the host, which are beneficial to health.

2) Postbiotics are deliberately inactivated microbial cells with or without metabolites or cell components that contribute to demonstrated health benefits.

Reading Material 3 Dietary Guidelines

Diet and nutrition are essential factors in promoting good health throughout life. Once the dietary intake data are collected, the next step is to determine the nutrient content of the diet and evaluate that information in terms of dietary standards or other reference points. This is commonly done using nutrient analysis software. Computer programs remove the tedium of looking up foods in tables of nutrient composition; large databases allow for simple access to food composition, and the computer does the math automatically. The main three methods for dietary evaluation are shown below.

Comparison to Dietary Standards

It is possible to compare a person's nutrient intake to dietary standards such as the RDA. Although this will give a qualitative idea of dietary adequacy, it cannot be considered a definitive evaluation of a person's diet because we don't know that individual's specific nutrient requirements. Comparisons of individual diets to RDA should be interpreted with caution.

Comparison to Food Guide Pyramid

Another type of dietary analysis compares a person's food intake to the Food Guide Pyramid. This involves categorizing foods into the various groups and determining the number of servings the subject has eaten. Evaluators often have trouble making these comparisons because many common foods (e.g., pizza, sandwiches, casseroles) contain servings or partial servings from multiple food groups.

Comparison to Dietary Guidelines

For a general picture of the subject's dietary habits, the evaluator can compare the person's diet to the Dietary Guidelines. While these evaluations usually are not specific, they give a general idea of whether the subject's diet is high or low in saturated fat, or whether the subject is eating enough fruits and vegetables.

About the Dietary Guidelines

Food and nutrition play a crucial role in health promotion and chronic disease prevention. Every 5 years, the U.S. Department of Health and Human Services (HHS) and the USDA publish the Dietary Guidelines for Americans, the nation's go-to source for nutrition advice.

The latest edition of the Dietary Guidelines reflects the current body of nutrition science, helps health professionals and policymakers guide Americans to make healthy food and beverage choices, and serves as the science-based foundation for vital nutrition policies and programs across the US.

The Dietary Guidelines provides evidence-based food and beverage recommendations for Americans across the lifespan. These recommendations aim to promote health and prevent chronic disease.

Public health agencies, health professionals, and educational institutions all rely on Dietary Guidelines recommendations and strategies.

The Dietary Guidelines has a significant impact on nutrition in the US because it:

(1) Forms the basis of federal nutrition policy and programs.

(2) Helps guide local, state, and national health promotion and disease prevention initiatives.

(3) Informs various organizations and industries, such as the food and beverage industry.

Current Dietary Guidelines

The Dietary Guidelines for Americans provides advice on what to eat and drink to meet nutrient needs, promote health, and prevent disease. It's developed and written for a professional audience, including policymakers, health care providers, nutrition educators, and federal nutrition program operators. This edition focuses on dietary pattern recommendations using a lifespan approach for all age groups.

The Dietary Guidelines provides a customizable framework for healthy eating that can be tailored and adapted to meet personal, cultural, and traditional preferences. People who work in federal agencies, public health, health care, education, and business all rely on the Dietary Guidelines when providing information on diet and health to the general public. This edition of the Dietary Guidelines includes:

(1) New recommendations for infants and toddlers.

(2) Expanded recommendations for people who are pregnant or breastfeeding.

(3) Updated recommendations for children and adolescents, adults, and older adults.

The information in the Dietary Guidelines is used to develop, implement, and evaluate federal food, nutrition, and health policies and programs. It also is the basis for federal nutrition education materials designed for the public and for the nutrition education components of USDA and HHS nutrition programs. State and local governments, schools, the food industry, other businesses, community groups, and media also use Dietary Guidelines information to develop programs, policies, and communication for the general public.

Notes

1) Although this will give a qualitative idea of dietary adequacy, it cannot be considered a definitive evaluation of a person's diet because we don't know that individual's specific nutrient requirements. Comparisons of individual diets to RDA should be interpreted with caution.

参考译文：虽然这可以定性说明膳食是否充足，但它不能被认为是对个人膳食的决定性评估，因为我们不知道这个人的具体营养需求。个人饮食与 RDA 的比较应谨慎解释。

句中"although"作连词放在句首，引导让步状语从句，可翻译为：虽然；即使。

2) For a general picture of the subject's dietary habits, the evaluator can compare the person's diet to the Dietary Guidelines. While these evaluations usually are not specific, they give a general idea of whether the subject's diet is high or low in saturated fat, or whether the subject is eating enough fruits and vegetables.

参考译文：要了解受试者饮食习惯的大致情况，评估者可以通过将受试者的饮食与膳食指南进行比较。虽然这些评估通常并不具体，但它们可以大致了解受试者的饮食中饱和脂肪的含量是高还是低，或者受试者是否吃了足够的水果和蔬菜。

句中"while"作连词，引导让步状语从句时，意思是"尽管，虽然"，一般用于句首。

Exercise

1. True or false.

1) Energy balance is the relationship between energy intake and energy control. (　　)

2) To lose weight, our dietary choices must provide fewer calories than we expend. (　　)

3) Everyone must eat according to the Recommended Dietary Guidelines to stay healthy. (　　)

2. Translate the following sentences into English.

1）膳食指南的制定，能够提供有科学依据的饮食建议，这些饮食建议有利于促进人类健康和预防慢性疾病。

2）膳食指南的信息可以用于食品的研发、应用和评估，也可以用于制订营养与健康政策和计划。

Chapter 3

Food Microbiology and Fermented Food

Lesson 1 Microorganism in Food

Although it is extremely difficult to find out the precise beginning of human awareness of the presence and role of microorganisms in foods, the available evidence indicates that this knowledge preceded the establishment of microbiology as a science. The microbial groups important in foods consist of several species and types of bacteria, yeasts, molds and viruses. They are important in food for their ability to cause foodborne diseases, food spoilage, or to produce food and food ingredients. Many bacterial species, some molds and viruses, but not yeasts, are able to cause foodborne disease. Most bacteria, molds, and yeasts, because of their ability to grow in foods, can potentially cause food spoilage. Several species of bacteria, molds, and yeasts are considered safe or food grade, or both, and are used to produce fermented foods. Among the four major groups, bacteria constitute the largest group. Because of their ubiquitous presence and rapid growth rate, even under conditions where yeasts and molds cannot grow, they are considered the most important in food spoilage and foodborne disease.

Bacteria are unicellular organisms that normally multiply by binary fission. Bacteria are classified partly by their appearance. However, to be able to see bacteria, they must be studied under a microscope at a magnification of about 1,000 times. Bacteria may also be stained and the most widely used method of staining bacteria was introduced by the Danish bacteriologist Gram and is called Gram staining. Bacteria are divided into two main groups according to their Gram stain characteristics: Gram negatives are red and Gram positives are blue. Bacteria come in many different shapes. However, three main characteristic shapes can be distinguished: spherical-, rod- and spiral-shaped.

Fungi are a group of micro-organisms that are frequently found in nature among plants, animals and human beings. Different species of fungi vary a great deal in structure and method of reproduction. Fungi may be round, oval or threadlike. The threads may form a network, visible to the naked eye, in the form of mould on food, for example. Fungi are divided into yeasts and molds.

Yeasts are single-cell organisms of spherical, elliptical or cylindrical shape. The size of yeast cells varies considerably. Brewer's yeast, *Saccharomyces cerevisiae*, has a diameter of 2-8 μm and a length of 3-15 μm. Some yeast cells of other species may be as large as 100 μm. Yeasts, like molds, have a more complex internal structure than bacteria. They contain cytoplasm and a clearly discernible nucleus surrounded by nuclear membrane. The cell is enclosed by a wall and a cell membrane, which is permeable to nutrients from the outside of the cell and waste products from the inside. The cell contains a vacuole that serves as storage space for reserve nutrition and for waste products before they are released from the cell. In the cytoplasm there is also a fine network of membranes named endoplasmic reticulum, mitochondria (where energy for cell growth is generated), as well as ribosomes.

Microbial food spoilage occurs as a consequence of either microbial growth in a food or release of microbial extracellular and intracellular enzymes in the food environment. Some of the detectable parameters associated with spoilage of different types of foods are changes in color, odor, and texture; formation of slime; accumulation of gas or foam; and accumulation of liquid. Spoilage by microbial growth occurs much faster than spoilage by microbial extra- or intracellular enzymes in the absence of viable microbial cells. Between initial production, such as harvesting of plant foods, and finial consumption, different methods are used to preserve the acceptance in qualities of foods, which include the reduction of microbial number and growth.

Beneficial microorganisms are used in foods in several ways. These include actively growing microbial cells, nongrowing microbial cells, and metabolic by-products and cellular components of microorganisms. An example of the use of growing microbial cells is the conversion of milk to yogurt by bacteria. Nongrowing cells of some bacteria are used to increase shelf life of refrigerated raw milk or raw meat. Many by products, such as lactic acid, acetic acid, some essential amino acids, and bacteriocins produced by different microorganisms, are used in many foods. Finally microbial cellular components, such as single-cell protein (SCP), dextran, cellulose, and many enzymes, are used in food for different purposes. These microorganisms or their by-products or cellular component must be safe, food grade, and approved by regulatory agencies.

Food fermentation involves a process in which raw materials are converted to fermented foods by the growth and metabolic activities of the desirable microorganisms. The microorganisms utilize some components present in the raw materials as substrates to generate energy and cellular components, to increase in population, and to produce many usable by-products (also called end-products) that are excreted in the environment. The unused components of the raw materials and the microbial by-products (and sometimes microbial cells) together constitute fermented foods. Worldwide, more than 3,500 types of fermented foods are produced. The old city civilizations, dating as far back as 5000-3000 B.C. in the Indus Valley, Mesopotamia, and Egypt, developed exceptional skills in the production of fermented foods from milk, fruits, cereal grains, and vegetables.

The basic principles developed by these ancient civilizations are used even today to produce many types of fermented foods by a process known as natural fermentation. In this method, either the desirable microbial population naturally present in the raw materials or some products containing the desirable microbes from a previous fermentation, are added to the raw materials. Then the fermentation conditions are set so as to favor growth of the desirable types but prevent or retard growth of undesirable types that could be present in the raw materials. In another type of fermentation, called controlled or pure culture fermentation, the microorganisms associated with fermentation of a food are first purified from the food, identified, and maintained in the laboratory. When required for the fermentation of a specific food, the microbial species associated with this fermentation are grown in large volume in the laboratory and then added to the raw materials in very large numbers. Then the fermentation conditions are set such that these microorganisms grow preferentially to produce a desired product. These microbial species, when used in controlled fermentation, are also referred to as starter cultures.

Fermentation can carry out by different kinds of microorganism, so can produce different products. Fermentation by lactic acid bacteria produces yogurt, pickles, olives, sausage, sour cream, cheddar cheese, coffee. Acetic acid bacteria produce cider and vinegar. Yeasts are involved in the production of beer, wine, whiskey, Chinese spirit, and bread. Molds can produce blue cheese.

Numerous food products owe their production and characteristics to the fermentative activities of microorganisms. Fermentation is the oldest form of food preservation. Many foods such as ripened cheeses, pickles, sauerkraut, and fermented sausages are preserved products in that their shelf life is extended considerably over that of the raw materials from which they are made. In addition to being made more shelves stable, all fermented foods have aroma and flavor characteristics that result directly or indirectly from the fermenting organisms. Or fermentation can alter the chemical characteristics of the food as in sugar to ethanol, ethanol to acetic acid, or sugar to lactic acid. Fermented food can be more nutritious than the unfermented foods from which they were derived. Fermentation microorganisms produce vitamins and growth factors in the food. They also may liberate nutrients locked in plant cells and structures by indigestible materials. Finally, fermentation can enzymatically split polymers like cellulose into simpler sugars that are digestible by humans.

The microbial ecology of food and related fermentations has been studied for many years. For example, when the natural raw materials are acidic and contain free sugars, yeasts grow readily, and the alcohol they produce restricts the activities of most other naturally contaminating organisms. If, on the other hand, the acidity of a plant product permits good bacterial growth and at the same time the product is high in simple sugars, lactic acid bacteria may be expected to grow, and the addition of low levels of NaCl will ensure their growth preferential to yeasts (as in sauerkraut fermentation). Products that contain polysaccharides but no significant levels of simple sugars are normally stable to the activities of yeasts and lactic acid bacteria due to the lack of amylase in most of these organisms. To effect fermentation, an exogenous source of saccharifying enzymes must be supplied. The use of barley malt in the brewing and distilling industries is an example of this. The fermentation of sugars to ethanol that results from malting is then carried out by yeasts. The use of koji in the fermentation of soybean products is another example of the way in which alcoholic and lactic acid fermentations may be carried out on products that have low levels of sugars but high levels of starches and proteins. Whereas the saccharifying enzymes of barley malt arise from germinating barley, the enzymes of koji are produced by *Aspergillus oryzae* growing on soaked or steamed rice or other cereals. The koji hydrolysates may be fermented by lactic acid bacteria and yeasts, as is the case for soy sauce, or the koji enzymes may act directly on soybeans in the production of products such as Japanese miso.

New Words

bacteria [bæk'tɪəriə]	*n.*	（复数）细菌
mold [məʊld]	*n.*	霉菌；模子
	vt.	形成；制模；发霉
	vi.	发霉；符合形状
virus ['vaɪrəs]	*n.*	病毒
yeast [jiːst]	*n.*	酵母
foodborne ['fuːdbɔːn]	*adj.*	食物传播的；食物传染的；食源性的
ubiquitous [juː'bɪkwɪtəs]	*adj.*	普遍存在的；无所不在的
fungi ['fʌŋgaɪ]	*n.*	真菌；菌类（fungus 的复数）
unicellular [juːnɪ'seljələ(r)]	*adj.*	单细胞的
binary fission		（原生动物、细胞等的）二分裂；二分体
microscope ['maɪkrəskəʊp]	*n.*	显微镜
Gram staining		革兰染色
oval ['əʊvl]	*adj.*	椭圆的；卵形的
	n.	椭圆形；卵形
threadlike ['θredlaɪk]	*adj.*	丝状的；细长的
spherical ['sferɪkl]	*adj.*	球形的；球面的
elliptical [ɪ'lɪptɪkl]	*adj.*	椭圆的
cylindrical [sə'lɪndrɪkl]	*adj.*	圆柱形的；圆柱体的
Saccharomyces cerevisiae		酿酒酵母
nucleus ['njuːkliəs]	*n.*	核；核心；原子核
cell membrane	*n.*	细胞膜
permeable ['pɜːmiəbl]	*adj.*	能透过的；有渗透性的
vacuole ['vækjuəʊl]	*n.*	液泡

endoplasmic reticulum	*n.*	内质网
mitochondria [ˌmaɪtəʊ'kɒndrɪə]	*n.*	线粒体（mitochondrion 的复数）
ribosome ['raɪbəsəʊm]	*n.*	核糖体
slime [slaɪm]	*n.*	黏液；烂泥
cellular ['seljələ(r)]	*adj.*	细胞的；多孔的
bacteriocin [bæktɪər'ɪəʊsɪn]	*n.*	细菌素
dextran ['dekstrən]	*n.*	右旋糖酐；葡萄聚糖
Mesopotamia [ˌmesəupə'teimjə]	*n.*	美索不达米亚（亚洲西南部）
retard [rɪ'tɑːd]	*vt.*	延迟；阻止
pickle ['pɪkl]	*n.*	泡菜；咸菜；腌渍物；各式腌菜
	v.	盐腌制；醋渍
sauerkraut ['saʊəkraʊt]	*n.*	一种德国泡菜
indigestible [ˌɪndɪ'dʒestəbl]	*adj.*	难消化的；难理解的
ecology [i'kɒlədʒi]	*n.*	生态学；社会生态学
exogenous [ek'sɒdʒənəs]	*adj.*	外生的；外因的；外成的
saccharify [sə'kærɪfaɪ]	*vt.*	使糖化
koji ['kəʊdʒɪ]	*n.*	日本酒曲；清酒曲
Aspergillus oryzae		米曲霉
hydrolysate [haɪ'drɒlɪseɪt]	*n.*	水解液；水解产物
miso ['miːsəʊ]	*n.*	味噌；日本豆面酱

Notes

1) Fungi may be round, oval or threadlike. The threads may form a network, visible to the naked eye, in the form of mould on food, for example.
   参考译文：真菌可能是圆形、椭圆形或丝状。例如，在食物表面肉眼可见的霉菌就是这些菌丝形成的网状物。
   句中“in the form of”含义是以…的形式。

2) Microbial food spoilage occurs as a consequence of either microbial growth in a food or release of microbial extracellular and intracellular enzymes in the food environment.
   参考译文：微生物导致的食品腐败是由于在食品中微生物的生长或是微生物的胞内和胞外酶释放到食品中的结果。
   句中“either…or..”表示或者…或者…，“microbial growth in a food”和“release of microbial extracellular and intracellular enzymes in the food environment”是并列关系。

3) The microorganisms utilize some components present in the raw materials as substrates to generate energy and cellular components, to increase in population, and to produce many usable by-products (also called end-products) that are excreted in the environment.

参考译文：微生物利用原料中一些成分产生能量和细胞组成成分，完成增殖，且生产分泌到环境中的有用副产物（也称终产物）。

句中“to generate energy and cellular components”“to increase in population”“to produce many usable by-products”是并列关系。“that”引出定语从句修饰“by-products”。

4) The basic principles developed by these ancient civilizations are used even today to produce many types of fermented foods by a process known as natural fermentation. In this method, either the desirable microbial population naturally present in the raw materials or some products containing the desirable microbes from a previous fermentation, are added to the raw materials. Then the fermentation conditions are set so as to favor growth of the desirable types but prevent or retard growth of undesirable types that could be present in the raw materials.

参考译文：这些古代文明所开发的基本原理甚至在今天还被用来生产许多种类的发酵食品，这个过程称为自然发酵。在这种方法中，不管是在原料中自然存在的有益微生物种群，或是包含来自前期发酵微生物的一些产物都添加到原料中。然后控制发酵条件使其适于有益微生物的生长，阻止或延迟存在于原料中的其他微生物的生长。

5) In addition to being made more shelf stable, all fermented foods have aroma and flavor characteristics that result directly or indirectly from the fermenting organisms. Or fermentation can alter the chemical characteristics of the food as in sugar to ethanol, ethanol to acetic acid, or sugar to lactic acid.

参考译文：除了获得更好的货架稳定性外，所有的发酵食品都会具有由发酵微生物直接或间接导致的风味和香气特性。发酵还能够改变食品的化学特性，正如由糖转化成乙醇，乙醇转化成乙酸，或是糖转化成乳酸。

6) Products that contain polysaccharides but no significant levels of simple sugars are normally stable to the activities of yeasts and lactic acid bacteria due to the lack of amylase in most of these organisms.

参考译文：含有多糖而非大量单糖的产品对于酵母和乳酸菌而言是稳定的，这是因为大多数这样的微生物缺少淀粉酶的缘故。

7) The use of koji in the fermentation of soybean products is another example of the way in which alcoholic and lactic acid fermentations may be carried out on products that have low levels of sugars but high levels of starches and proteins.

参考译文：在豆制品发酵中酒曲的使用就是这种方式的另一个实例，即酒精和乳酸发酵可以利用含低水平糖而高水平淀粉和蛋白质的原料来进行。

Lesson 2 Alcoholic Beverages

Alcoholic beverages have one point in common. They all depend on the process of fermentation — the conversion of hexose sugar into alcohol and carbon dioxide. This is indeed a very important process and is basic to all of the industries involved. Alcoholic fermentation is an anaerobic process carried on by living yeast cells. The cells absorb the simple sugars, are broken down in a series of successive changes in which action by oxidizing and reducing enzymes within the cell takes place. The final result is the formation of ethyl alcohol and carbon dioxide accompanied by the liberation of some energy in the form of heat.

Over 96% of fermentation ethanol is produced using strains of *Saccharomyces cerevisiae* or species related to it, particularly *Saccharomyces uvarum*. Ethanol is produced by the embden-meyerhof-parnas (EMP) pathway. Pyruvate, produced during glycolysis is converted to acetaldehyde and ethanol. The overall effect is summarized as follows

$$\text{Glucose} + 2\text{ADP} \rightarrow 2\text{Ethanol} + 2CO_2 + 2\text{ATP}$$

According to the equation theoretical yields from 1 g glucose are 0.51 g ethanol and 0.49 g CO_2. However, in practice, the practical yields of alcohol from this reaction are from 90% to 95% of theory, because some of the glucose is used by the yeast cells for growth and some is converted into small quantities of other carbon compounds. Many enzymes are necessary to bring about the changes in alcoholic fermentation. This process is a complex system which explains why such compounds as glycerol, lactic acid, and acetaldehyde build up as by-product.

Brewing yeast can utilize some simple sugars, such as glucose, fructose, sucrose, maltose and maltotriose, which are called fermentable sugar. Other long-chain polysaccharides are called non-fermentable sugar because they cannot be utilized by yeast directly. Use of starch-based materials for alcoholic fermentations requires addition of exogenous enzymes such as α-and β-amylase of malt or microbial enzymes such as α-amylase, amyloglucosidase (glucoamylase) and pullulanase. The enzymes can break down the long chain of starch and finally produce shorter chain dextrin and simple sugars which can be used by yeast. For sugar-containing raw materials, such as grape, the major sugars are glucose and fructose and since *Saccharomyces cerevisiae* can metabolizes directly, the step of enzyme hydrolyzing is not needed.

Through alcoholic fermentation, the alcoholic beverages can be produced from a range of raw materials but especially from cereals, fruits and sugar crops. They include non-distilled beverages such as beers, wines, ciders, and sake. Distilled beverages such as whisky, rum, brandy, vodka and gin are produced by distillation from neutral spirits.

Non-distilled Beverages

Because the microorganisms cannot produce so high alcohol content, only beer, wine, cider, and sake which are low alcohol content beverages can be produced no distillation process.

Beer is a beverage whose history can be traced back to between 6,000 and 8,000 years. The basic ingredients for most beers are malted barley, water, hops, and yeast. During beer manufacture process, there exists the stage—mashing, in which action of enzyme and substrate occur and suitable growth medium—wort containing fermentable sugar is prepared. Compared to most other alcoholic beverages, beer is relatively low in alcohol. The highest average strength of beer in any country worldwide is 5.1% by volume and the lowest is 3.9%. There are two general types of beer, Ales and Lagers. Traditionally, top-fermenting yeasts, which ferment at 15-22℃ and tend to rise to the surface towards the end of fermentation and can therefore be skimmed off, were used in ale production. Bottom-fermenting yeasts, which ferment in the range 8-15℃ and sink to the bottom towards the end of fermentation, were generally used in lager production. Hops and malt have a major impact on the flavor, aroma, and color properties of the beer. Thus, beers are often classified based on their malt and hop type and content. In recent years, some new and innovative products and processes are developed, such as dry beer, ice beer, wheat beer, etc.

Wine is legally defined as "the product which is recovered exclusively by complete or partial alcoholic fermentation of the fresh, treated grapes or the grape must". A variety of wines exist. Still wine, which do not retain the carbon dioxide of the fermentation process, are the usual table wines, and are available as red, white, and rose. The red wines are colored by the pigments found in the skins of the grape. Rose wines are made by removing the skins during fermentation to reduce the amount of color extraction. White wines are made from white grapes or from red grapes which the juice has been pressed before fermentation. The alcohol by volume of still wine is typically in the range 11%-15%. Fortified wines are produced by addition of distilled spirit to still wines to raise the alcohol content to 15%-22% by volume. Notable products include Sherries, Port and Madeira wines. Champagne, which is one of the most famous sparkling wine, undergoes a second alcoholic fermentation without the escape of the formed carbon dioxide. If carbon dioxide is artificially introduced, simple sparkling wine is obtained. With white wines, lower fermentation temperatures produce fresher and fruitier wines, and the risk of bacterial infection and resultant volatile acid production is reduced. Higher temperatures of 22-30℃ are used for production of red wines, fermented on the skins, leading to increased colour extraction and production of a rich aroma.

Distilled Spirits

There are many kinds of distilled spirits in the world, and the raw material of each is different too. Whisky and rum are produced from fermented cereals and molasses, respectively, while brandy is produced by distillation of wine. Other distilled beverages, such as vodka and gin, are produced from neutral spirits obtained by distillation of fermented molasses, grain, potato or whey. All distilled alcohols are made in basically the same way.

Fermentation: fruits, grains, or other sources of carbohydrates are fermented with yeasts to make a liquid with moderate alcohol content, from 5% to 12% by volume.

Distillation: this liquid is heated in a chamber that collects the alcohol and aroma rich vapors as they escape from the boiling liquid, and then passes them across cooler metal surfaces, where the vapors condense and are collected as a separated liquid.

Modification: the concentrated alcoholic liquid is then modified in various ways for consumption. It may be flavored with herbs or spices or aged in wood barrels. The alcohol content is usually adjusted with the addition of water before it's bottled for sale.

Chinese liquor is a kind of liquor that has been passed down through generations in China, and always has played significant roles during its over 5,000 year history. Nowadays, white spirits of various brands are made from grains, tubers, and other starch- or sugar-containing materials. The usual alcohol content ranges from 40% to 65% (*V/V*). The distinguishing features of each brand of white spirits are determined by the variety of qu used, the fermentation processes, and its characteristic flavors. With follow-up research and summary, the traditional processing techniques have been renovated, from workshop processing to industrialized manufacture, from manual operation to semi-mechanized processing, from oral instruction to teaching by written materials.

New Words

anaerobic [æneə'rəʊbɪk]	*adj.*	厌氧的，厌气的
hexose ['heksəʊs]	*n.*	己糖
Saccharomyces uvarum		葡萄汁酵母
embden-meyerhof-parnas (EMP) pathway		糖酵解途径
pyruvate [paɪ'ruːveɪt]	*n.*	丙酮酸盐；丙酮酸酯
glycolysis [glaɪ'kɒlɪsɪs]	*n.*	糖酵解
acetaldehyde [ˌæsɪ'tældəhaɪd]	*n.*	乙醛；醋醛
glycerol ['glɪsərɒl]	*n.*	甘油；丙三醇
non-fermentable sugar		非发酵性糖
amyloglucosidase [eɪmɪlɒgluː 'kəʊsaɪdeɪz]	*n.*	淀粉葡萄糖苷酶
pullulanase [pʌl'jʊlænəs]	*n.*	支链淀粉酶
hydrolyze ['haɪdrəlaɪz]	*vi.*	水解
	vt.	使水解
distilled beverages		蒸馏酒
cider ['saɪdə(r)]	*n.*	苹果酒；苹果汁
sake ['sɑːki]	*n.*	日本米酒，清酒
whisky ['wɪski]	*n.*	威士忌酒
	adj.	威士忌酒的
rum [rʌm]	*n.*	朗姆酒；甜酒（用甘蔗或糖蜜等酿制的一种甜酒）

	adj.	古怪的；危险的；困难的
molasses [mə'læsɪz]	*n.*	糖蜜；糖浆
brandy ['brændi]	*n.*	白兰地酒；（夹心）糖果
	v.	以白兰地酒调制
vodka ['vɒdkə]	*n.*	伏特加酒
gin [dʒin]	*n.*	杜松子酒；弹棉机；轧花机；陷阱；网
	v.	用轧棉机去籽；用陷阱（网）捕捉
malt [mɔːlt]	*n.*	麦芽；麦芽酒
	adj.	麦芽的
	vt.	用麦芽处理
barley ['bɑːli]	*n.*	大麦
hops [hɒps]	*n.*	啤酒花（名词 hop 的复数形式）
top-fermenting yeast		上面发酵酵母
bottom-fermenting yeast		下面发酵酵母
ale [eɪl]	*n.*	爱尔啤酒；麦芽酒
lager ['lɑːgə(r)]	*n.*	拉格啤酒；窖藏啤酒
pigment ['pɪgmənt]	*n.*	色素；颜料
	v.	给…着色
fortified ['fɔːtɪfaɪd]	*adj.*	加强的
	n.	强化酒
sherry ['ʃeri]	*n.*	雪利酒（西班牙产的一种烈性白葡萄酒）；葡萄酒
Madeira [mə'dɪərə]	*n.*	马德拉群岛（大西洋的群岛名）；马得拉白葡萄酒
Champagne [ʃæm'peɪn]	*n.*	香槟酒；香槟酒色
sparkling ['spɑːklɪŋ]	*adj.*	闪闪发光的；闪烁的；起泡沫的
tuber ['tjuːbə(r)]	*n.*	［植］块茎；隆起
whey [weɪ]	*n.*	乳清；乳浆
chamber ['tʃeɪmbə(r)]	*n.*	（身体或器官内的）室，膛；房间；会所
	adj.	室内的；私人的

Notes

1) Use of starch-based materials for alcoholic fermentations requires addition of exogenous enzymes such as α-and β-amylase of malt or microbial enzymes such as α-amylase,

amyloglucosidase (glucoamylase) and pullulanase. The enzymes can break down the longchain of starch and finally produce shorter chain dextrin and simple sugars which can be used by yeast.

参考译文：使用淀粉质原料进行酒精发酵需要添加外加酶，例如麦芽中的 α-和 β-淀粉酶或微生物酶，如 α-淀粉酶、淀粉葡萄糖苷酶和支链淀粉酶。这些酶能分解长链的淀粉，最终产生短链的糊精和单糖，从而被酵母利用。

句中“starch-based materials”指的是淀粉质原料。“which”引出定语从句。

2) During beer manufacture process, there exists the stage—mashing, in which action of enzyme and substrate occur and suitable growth medium—wort containing fermentable sugar is prepared.

参考译文：在啤酒生产过程中存在糖化阶段，在此阶段酶和底物发生作用，包含可发酵性糖的适宜培养基——麦汁被制得。

3) Champagne, which is one of the most famous sparkling wine, undergoes a second alcoholic fermentation without the escape of the formed carbon dioxide. If carbon dioxide is artificially introduced, simple sparkling wine is obtained.

参考译文：最著名的起泡葡萄酒之一——香槟要进行二次酒精发酵，此过程中形成的二氧化碳没有溢出。如果人工添加二氧化碳可制得简易起泡葡萄酒。

4) Other distilled beverages, such as vodka and gin, are produced from neutral spirits obtained by distillation of fermented molasses, grain, potato or whey.

参考译文：其他的蒸馏酒（如伏特加和金酒），通过对发酵糖蜜、谷物或乳清进行蒸馏获得中性烈酒来进行生产。

句中“obtained by distillation of fermented molasses, grain, potato or whey”是过去分词作为定语，修饰“neutral spirits”。

5) This liquid is heated in a chamber that collects the alcohol and aroma rich vapors as they escape from the boiling liquid, and then passes them across cooler metal surfaces, where the vapors condense and are collected as a separated liquid.

参考译文：液体在一个小室（蒸馏室）内加热沸腾，收集富含酒精和香气成分的蒸汽，然后使其通过冷却器的金属表面，在此蒸汽被冷凝成液体后收集。

句中“where”引出状语从句。

6) The distinguishing features of each brand of white spirits are determined by the variety of qu used, the fermentation processes, and its characteristic flavors.

参考译文：每个不同品牌白酒的特征由使用曲的种类、发酵流程和风味特征来决定。

Lesson 3　Fermented Milk

Fermentation is one of the oldest methods practiced by human beings for the transformation of milk into products with an extended shelf life. The exact origin of the making of fermented milks is difficult to establish, but it could date from some 15,000 years ago as the way of life of human beings changed from being food gatherer to food producer. Cultured milk originates from the Near East and subsequently became popular in Eastern and Central Europe. The first example of cultured milk was presumably produced accidentally by nomads. This milk turned sour and coagulated under the influence of certain microorganisms. As luck would have it, the bacteria were of the harmless, acidifying type and were not toxin-producing organisms.

The generic name of fermented milk is derived from the fact that the milk for the product is inoculated with a starter culture which converts part of the lactose to lactic acid. Carbon dioxide, acetic acid, diacetyl, acetaldehyde and several other substances are formed in the conversion process, and these give the products their characteristic fresh taste and aroma. Around 400 generic names are applied to the traditional and industrialized fermented milk products manufactured throughout the world. Taking into account the type of milk used, the microbial species which dominate the flora and their principal metabolic products, fermented milks can be divided into three broad categories: lactic fermentations (such as buttermilk and yoghurt), yeast-lactic fermentations (Kefir, Koumiss) and mould-lactic fermentations.

The conversion of lactose into lactic acid has a preservative effect on milk. The low pH of fermented milk inhibits the growth of putrefactive bacteria and other detrimental organisms, thereby prolonging the shelf life of the product. On the other hand, acidified milk is a very favourable environment for yeasts and molds, which cause off-flavours if allowed to infect the products.

The digestive systems of some people lack the lactase enzyme. As a result, lactose is not broken down in the digestive process into simpler types of sugars. These people can consume only very small volumes of ordinary milk. They can, however, consume fermented milk, in which the lactose is already partly broken down by the bacterial enzymes.

In the production of fermented milk, the best possible growth conditions must be created for the starter culture. These are achieved by heat treatment of the milk to destroy any competing micro-organisms. In addition, the milk must be held at the optimum temperature for the relevant starter culture. When the best possible flavour and aroma have been achieved, the cultured milk must be cooled quickly, to stop the fermentation process. If the fermentation time is too long or too short, the flavour will be impaired and the consistency wrong.

In addition to flavour and aroma, correct appearance and consistency are important features. These are determined by the choice of pre-processing parameters. Adequate heat treatment and homogenization of the milk, sometimes combined with methods to increase the milk solid non-fat (MSNF) content, as for milk intended for yoghurt, are essential for the construction of the coagulum during the incubation period.

The yeast-lactic fermentations product, Kefir is one of the oldest fermented milk products. It originates from the Caucasus region. The raw material is milk from goats, sheep or cows. Kefir is produced in many countries, although the largest quantity is consumed in Russia. Kefir should be viscous and homogenous with a shiny surface. The taste should be fresh and acid, with a slight flavour of yeast. The pH of the product is usually 4.3-4.4. A special culture, known as Kefir grain, is used for the production of Kefir. The grains consist of proteins, polysaccharides and a mixture of several types of micro-organisms, such as yeasts and aroma and lactic acid forming bacteria. The yeasts represent about 5%-10% of the total microflora.

Fermented buttermilk is manufactured on many markets in order to overcome problems such as off-flavours and short shelf life. The raw material can be sweet buttermilk from the manufacture of butter based on sweet cream, skim milk or low-fat milk. In all cases the raw material is heat treated at 90-95℃ for about 5 minutes before being cooled to inoculation temperature. Ordinary lactic acid bacteria are most commonly used. Buttermilk may also be flavoured with fruit concentrate.

Yogurt is another well-known cultured milk fermented using a symbiotic mixture of *Streptococcus thermophilus* (coccus) and *Lactobacillus bulgaricus* (rod) as starter. Consumption of yoghurt is highest in countries around the Mediterranean, in Asia and in Central Europe. The consistency, flavour and aroma vary from one district to another. In some areas, yoghurt is produced in the form of a highly viscous liquid, whereas in other countries it is in the form of a softer gel. Yoghurt is also produced in frozen form as a dessert, or as a drink. The flavour and aroma of yoghurt differ from those of other acidified products, and the volatile aromatic substances include small quantities of acetic acid and acetaldehyde. In most of Western and Eastern Europe, yogurt is made from cow's milk. In the Baltic regions and the Middle East, milk from goats is also used and in the Indian subcontinent milk from buffaloes, in addition to cow's milk, is used in yogurt production.

According to Food and Drug Administration (FDA), yogurt is the food produced by culturing the following, cream, milk, partially skimmed milk or skim milk either alone or in combination, with a characterizing bacterial culture that contains the lactic acid bacteria. Additionally, the regulation specifies that yogurt before addition of bulky flavours contains not less than 3.25% milk fat and not less than 8.25% MSNF and has a titrable acidity not less than 0.9%, expressed as lactic acid. Three categories of the product recognized according to fat content are yogurt, low fat yogurt and non-fat yogurt. A product to be labeled yogurt should meet all the aforementioned criteria. Low fat yogurt should meet all the criteria specified for yogurt, but the fat content should range between 0.5% and 2.0%. Non-fat yogurt should contain less than 0.5% milk fat. Moreover, yogurt can be classified as follows:

• Set type: incubated and cooled in the package.

• Stirred type: incubated in tanks and cooled before packing.

• Drinking type: similar to stirred type, but the coagulum is broken down to a liquid before being packed.

• Frozen type: incubated in tanks and frozen like ice cream.

• Concentrated: incubated in tanks, concentrated and cooled before being packed. This type is sometimes called strained yoghurt, sometimes labneh.

The latest years there has been increased focus on functional foods. Within this category certain types of lactic acid bacteria plays a large role. For a number of years, it has been known that a certain type of cultured milk called Langfil has been used to heal wounds and treat vaginal fungus infections. However, studies of lactic acid bacteria and their importance to health can be traced back to the beginning of the twentieth century. Many people in Russian consumed a great deal of yoghurt and lived for a long time.

Thus, lactic acid bacteria may have a great potential for promoting the health of both human beings and animals. It is therefore important that sufficient resources are invested in this field in the near future, both to find new interesting health effects of lactic acid bacteria and to compile scientific documentation.

New Words

coagulate [kəʊ'ægjuleɪt]	*vt.*	使…凝结
	vi.	凝结
inoculate [ɪ'nɒkjuleɪt]	*vt.*	接种；嫁接；灌输
diacetyl [daɪə'siːtɪl]	*n.*	双乙酰
buttermilk ['bʌtəmɪlk]	*n.*	酪乳；白脱牛奶；脱脂乳
Kefir ['kefə]	*n.*	（俄）（由牛奶发酵而成的）克非尔
koumiss ['kuːmɪs]	*n.*	酸马奶；马奶酒（等于 kumiss）
putrefactive [ˌpjuːtrɪ'fæktɪv]	*adj.*	腐败的；腐烂的；易腐败的
digestive [daɪ'dʒestɪv]	*adj.*	消化的；助消化的
	n.	消化药
homogenization [həumədʒənaɪ'zeɪʃən]	*n.*	均化；均化作用
MSNF (milk solid non-fat)		非脂乳固体
coagulum [kəʊ'ægjʊləm]	*n.*	凝结物；凝固物；凝块
Caucasus ['kɔːkəsəs]	*n.*	高加索（俄罗斯南部地区，位于黑海和里海之间）
microflora [ˌmaɪkrəʊ'flɒrə]	*n.*	微生物区系；微生物群落
off-flavours		异常风味
Streptococcus thermophilus		嗜热链球菌
coccus ['kɒkəs]	*n.*	球菌；小干果
Lactobacillus bulgaricus		保加利亚乳杆菌
rod [rɒd]	*n.*	棒；惩罚；枝条
Mediterranean [ˌmedɪtə'reɪniən]	*n.*	地中海
	adj.	地中海的
volatile ['vɒlətaɪl]	*adj.*	挥发性的；不稳定的
	n.	挥发物；有翅的动物

buffaloe ['bʌfələu]	*n*.	水牛；野牛
skim [skɪm]	*vt*.	略读；撇去…的浮物
	vi.	浏览
	adj.	脱脂的
titrable acidity		可滴定酸度
set yoghurt		凝固酸奶
stirred yoghurt		搅拌酸奶
strained yoghurt		脱乳清酸奶
labneh		浓缩酸奶
Langfil		瑞典酸奶

Notes

1) The generic name of fermented milk is derived from the fact that the milk for the product is inoculated with a starter culture which converts part of the lactose to lactic acid. Carbon dioxide, acetic acid, diacetyl, acetaldehyde and several other substances are formed in the conversion process, and these give the products their characteristic fresh taste and aroma.
参考译文：发酵乳的通用名称源于其生产过程，即在牛奶中接种发酵剂，从而将乳糖转化为乳酸而制得。在转化过程中，二氧化碳、乙酸、双乙酰、乙醛和其他的一些物质生成，给予产品特有的新鲜味道和香气。
句中“derived from”由…而来。“that”引出从句修饰“the fact”。“which”引出定语从句修饰“starter culture”。

2) The low pH of fermented milk inhibits the growth of putrefactive bacteria and other detrimental organisms, thereby prolonging the shelf life of the product. On the other hand, acidified milk is a very favourable environment for yeasts and moulds, which cause off-flavours if allowed to infect the products.
参考译文：发酵乳的低pH值抑制腐败细菌和其他有害微生物的生长，从而延长产品的保质期。另一方面，酸化乳给酵母和霉菌提供了一个适宜的环境，如果其污染产品则会引起产品风味异常。

3) Adequate heat treatment and homogenisation of the milk, sometimes combined with methods to increase the milk solid non-fat (MSNF) content, as for milk intended for yoghurt, are essential for the construction of the coagulum during the incubation period.
参考译文：将准备做酸奶的牛奶进行适当的热处理和均质处理，有时结合增加非脂固形物含量的方法对于培养过程中凝结物的结构是必要的。

4) Yogurt is another well-known cultured milk fermented using a symbiotic mixture of *Streptococcus thermophilus* (coccus) and *Lactobacillus bulgaricus* (rod) as starter.
参考译文：酸奶是另一个众所周知的发酵乳，是用嗜热链球菌（球菌）和保加利亚乳杆菌（杆菌）的共生混合物作为发酵剂的。

句中“starter”的含义是发酵剂。“symbiotic mixture”含义是共生的混合物。

5) According to Food and Drug Administration (FDA), yogurt is the food produced by culturing the following, cream, milk, partially skimed milk or skim milk either alone or in combination, with a characterizing bacterial culture that contains the lactic acid bacteria.

参考译文：根据美国食品和药品管理局的规定，酸奶是由奶油、牛奶、部分脱脂乳和脱脂乳作为原料单独或结合使用，并利用包含有乳酸菌的特殊细菌培养物进行培养而制得的食品。

句中“FDA”即美国食品和药品管理局。

Lesson 4 Fermented Meat Products

One of the most important prerequisites for the development of civilization was to devise methods on how to preserve foods for storage and transport in order to meet the need for food for increasing concentration in communities. Drying of foodstuffs was probably the first development in this direction, followed by smoking, which in many cases was a natural consequence; drying often was accelerated by hanging up the raw material near the open fire. Nobody had the idea that other processes such as enzymatic breakdown and product changes caused by microorganisms or endogenous enzymes could also be the reason for extension of product life.

However, the transformation of raw materials to more or less stable foods by drying and fermentation is well known in many ancient cultures and used for many different foods. Actually, drying as an initial step of preservation, or the “wet way”, such as steeping often followed by heating, is well known in beer production, for example, where the breakdown of raw materials takes place because the enzymes are brought into a suitable environment regarding pH and water activity.

The very word “fermentation”, derived from the Latin, means among other things to simmer or bubble, or leaven as a process, was probably not well understood, except that the effect was certainly used when it came to baking, wine making, beer brewing, and production of dairy products or certain meat products. However, because it was poorly understood, this frequently resulted in faulty production. This happens today. A production manager in a factory for producing Parma Hams about 25 years admitted that up to 25% of the raw material never reached a stage qualifying for the stamp of Parma Ham because it was putrid before it was ready for sale.

Today, some uncertainties still exist regarding how to define different fermented meats. Thus, fermented food is usually described in three categories: (a) those in which microorganisms play no or little part, such as tea fermentation; (b) those in which microbial growth, though an essential feature, does not involve fermentative metabolism, such as the production of tempeh; (c) the true fermentations, which produce, for example, lactic acid in products such as salami. The history of fermented meat products is very old. It includes both raw, dried ham and sausages, because those categories of products alter qualities during production and storage due to fermentation.

Raw Cured Ham

As earlier mentioned, drying, as a means of extending shelf life of food, was probably the earliest technique employed by man. This method, however, could be used only if the food was very easy to dry or semidry to begin with, for example, seeds, or if the climatic conditions were suitable, such as carne seca, which are South American air-dried meat specialities. In most other parts of the world, some kind of salting had to be used in addition to prevent rot of the raw material. The combination was well known in several ancient civilizations throughout the world—Rome and even Gaul and China. It is a method used today for the production of traditional dried hams.

Several attempts have been made to speed up the process. Thus, Puolanne reports on the pickle injection of raw ham. The method is also used in several other countries today. It reduces the production time considerately, from up to a year to a couple of weeks. Puolanne further reports that these kinds of ham have much less aroma than hams manufactured the traditional way. This observation is reported from several countries. The attempts to hasten the production and improve the safety precautions in ham production have led to the development of inoculation of microorganisms in dry ham.

To store dry ham at a higher temperature or to expose it at a higher temperature for a short time during the later part of manufacture is also used in some cases. Puolanne thus reports on a Finnish-made dry-cured ham, called a sauna ham, which was manufactured earlier. These hams were made as follows: after slaughter, which took place in the autumn, the hams were dry-salted and placed in wooden barrels for 1-4 months. Then they were hot-smoked at approximately 40-70℃ or applied a cold smoke at approximately 15-20℃ for several days.

Smoking, a cold smoke at 20-25℃, has traditionally been used in many countries. Various types of German dry hams are smoked, whereas the Parma type ham is never smoked; neither is the Iberian ham.

Dry Sausage

Cutting up meat in smaller pieces to enable a more uniform distribution of the salt and other ingredients has been a well-known technique since the Greek and Roman days. However, to comminute meat also opened it up for a thorough contamination, not only of a desired flora, but also of bacteria, which sometimes could be pathogens. It was especially important to exclude air.

This often caused faulty productions, especially blowing and discolorations. Before artificial refrigeration and control of humidity was available, it was almost impossible to produce dry sausages of satisfactory quality. Because of these difficulties, and because of the demand for a dry sausage, which had at least some keep ability, the summer sausage was developed. The name summer sausage is still used in some countries; in others, it is not. Thus, in the united states, dry and/or semidry sausages are often called summer sausages, but the term certainly covers a lot of names, which in other parts of the world would be called typical fermented sausages, meant to have shelf lives far beyond a few weeks. Rather, the classical dry sausage will have a keep ability of a year or more.

The modern way of production of dry sausage with the use of starter cultures and full control of temperatures and humidity has completely changed the manufacture of today's fermented sausage.

The tradition for consuming fermented sausages in Europe varies, and it relates traditionally to the area. Thus, probably because of the climate, fermented sausages were practically not produced or consumed on the British Isles, and in Ireland in earlier times, only whole, dried meats such as ham were known. On the European continent, fermented sausages were produced abundantly, a tradition starting in Southern Europe.

Starter Cultures

The use of starter cultures in various types of foods has been used for a long time, probably for as long as fermentation has been used by man; but it was used simply as a kind of back-slopping of remains of earlier production charges. No doubt the dairy industry has been using back-slopping purposely, for example, for making junket and other kinds of fermented milk long before anybody knew anything about bacteria. The process was derived from simple experience in the home. As Pederson describes it, "the mothers observed that when milk soured with a smooth curd and pleasing acid odor, the milk could be consumed without causing distress or illness. When instead, the milk had an unpleasing odor and was spoiled or gassy or otherwise exhibited unsatisfactory character, the infant sometimes became ill after consuming such milk". Pasteurization made even back-slopping much safer, but the use of defined starter cultures made the production a much more foolproof processes. With meat, it was different. Microorganisms were added through natural contamination, and to this very day back-slopping are still commonly used in many places and regions of the world.

It is stated that many productions are still made successfully without addition of starter cultures or reinoculation—back-slopping of finished sausages. This is due to sausage makers being able to design formulations and ripening conditions that favor the desired microorganisms so strongly that the products are safe and palatable, even if only few *lactobacilli* and *micrococci* (i.e., *Kocuria* and *staphyloccoci*) are present in the fresh mixture. No doubt the interest in the use of starter cultures arose parallel with the trend toward industrial production, short ripening times, and standardization of the mode in which the sausages were made. It was also very helpful that contemporary refrigeration and air-conditioned facilities became available. However, due to the increasing use of starter cultures, which had taken place with such great effect in the dairy industries, attempts were made to develop starter cultures for other foods, such as meats.

It is impossible to cover the whole history of fermented meats easily, but going through some of the major epochs, the technical discoveries have resulted in far more rational procedures and thus changed many fermented meat products. These items sometimes were successful, but far from always, because many of the secrets of fermentation were unveiled. On the other hand, many discoveries still wait to be made, especially as far as ways in which more refined aroma and taste compounds are formed during processing. No doubt there is still much to discover in the future about the mechanisms and the control of volatile development.

New Words

prerequisite [ˌpriːˈrekwəzɪt]	*n.*	先决条件；前提；必备的事物
	adj.	作为前提的；必备的
concentration [ˌkɒnsenˈtreɪʃən]	*n.*	集中；聚集；专注；浓度；密度；浓缩；浓缩物
endogenous [enˈdɒdʒənəs]	*adj.*	内生的；内源的
transformation [ˌtrænsfəˈmeɪʃn]	*n.*	转化；变形；改造；转变
metabolism [məˈtæbəlɪzəm]	*n.*	代谢；新陈代谢
sausage [ˈsɒsɪdʒ]	*n.*	香肠；腊肠
contamination [kənˌtæmɪˈneɪʃn]	*n.*	弄脏；污染；杂质；污染物
pathogen [ˈpæθədʒən]	*n.*	病原体；致病菌；病原菌
reinoculation [reɪnɒkjʊˈleɪʃn]	*n.*	再接种
contemporary [kənˈtemprəri]	*adj.*	同时代存在的；同年龄的；当代的
	n.	同代人
refrigeration [rɪˌfrɪdʒəˈreɪʃn]	*n.*	制冷；冷藏；制冷
aroma [əˈrəʊmə]	*n.*	香味；芳香；气味；（艺术品等的）格调；韵味
starter [ˈstɑːtə(r)]	*n.*	起始者，开创者；起始物；发酵剂

Notes

1) To store dry ham at a higher temperature or to expose it at a higher temperature for a short time during the later part of manufacture is also used in some cases.

参考译文：在某些情况下，制造干香肠的后期采用在较高温度下贮藏或者将其暂时暴露在高温下的方法。

2) Before artificial refrigeration and control of humidity was available, it was almost impossible to produce dry sausages of satisfactory quality.

参考译文：没有有效的人工制冷和湿度控制，几乎是不可能生产出质量让人满意的干香肠的。

3) This is due to sausage makers being able to design formulations and ripening conditions that favor the desired microorganisms so strongly that the products are safe and palatable, even if only few *lactobacilli* and *micrococci* (i.e., *Kocuria* and *staphyloccoci*) are present in the fresh mixture.

参考译文：这是由于香肠制造者能够设计出有利于目标微生物生长的配方和成熟环境，以至于即使在新鲜混合物中含有少量的乳酸杆菌和微球菌，也能保证产品的安全和适口。

Lesson 5 Fermented Vegetable Products

Introduction

Raw fruits and vegetables constitute foods of high nutritional and functional value with appreciated health-promoting properties. However, due to the short shelf life they have, a large amount of them is discarded as waste generating large economic losses and the accumulation of organic waste. The elaboration of fermented plant-based foods and beverages constitute an alternative for its sustainable use, also transforming them in carriers for delivery of potential probiotics of value to consumers suffering from allergy to milk proteins or lactose intolerance. The utilization of a controlled manufacturing process using beneficial autochthonous microorganisms instead of traditionally used spontaneous fermentation is recommended to obtain foods with the desirable nutritional and functional, as well as sensory and technological properties. Thus, fruits and vegetables can be transformed into fermented foods with specific properties targeted to improve specific pathologies or human health generally. Besides, enhancing the functional features and health benefits of fermented fruit or vegetable foods, it is of the utmost importance that the selection of new autochthonous microorganisms is aimed at improving the sensory quality of the fermented product, and therefore the acceptability by consumers. These aspects can be satisfied through the selection of microorganisms producing key aroma compounds and exopolysaccharides to improve flavor and consistency and overall attractiveness of the final product.

Fermentation of vegetables as a way of preservation and the consumption of fermented vegetables have a long history, practically going along with the development of human civilization. A clear example is the sauerkraut (sour cabbage) consumed by the inhabitants of the Roman Empire or the traditional sauerkrauts from Asia, such as the Chinese pao cai or the Korean “Kimchi”. Other ancient typical fermented vegetable foods from Asia include the Chinese “Jiangshui” (fermented celery, Chinese cabbage, mustard, radish sprouts, and potherbs) or fermented soybeans, such as Japanese “Miso” and “Natto” or the Indonesian “Tempeh”. Fermented vegetables have always received attention concerning their health benefits, which is continuously increasing, especially in “health-conscious” developed countries. Vegetables and fruits are per se sources of nutrients, such as vitamins, minerals, prebiotic fibers and bioactive phytochemicals, including phenolic acids, flavonoids, phytoestrogens and bioactive peptides. A growing number of scientific research have demonstrated the positive relationship between the consumption of a diet rich in vegetables and reducing the risks of age-related diseases, inflammatory pathologies, or metabolic disorders. Fermentation of fruits and vegetables, which is the main and simple alternative to heat or chemicals for guarantee the safety of vegetable products, may also lead to positive changes in the concentration or composition of vitamins, amino acids, bioactive peptides, or phytochemicals. Fermentation may also contribute to improving the bioavailability of these compounds and sensory attributes of vegetables and fruits. It is due to these considerations, that fermentation represents significant biotechnology to improve the safety, as well as the health-promoting properties and general attractiveness of fruits and vegetables.

Vegetables and fruits have a microbial population whose composition depends on the characteristics of each vegetable matrix as well as geographical origin. This microbiota, composed mainly by beneficial microorganisms, including yeasts (*Saccharomyces*, *Pichia*, *Candida*, *Torulaspora* genera), fungi (*Rhizopus* spp.), and aerobic (*Bacillus* spp. and *Acetobacter* spp.) and anaerobic (lactic acid bacteria) bacteria, is usually responsible for the spontaneous fermentation of raw vegetables and fruits, contributing to their preservation and stability. Under favorable conditions, fermentation of vegetables and fruits occurs mainly by the instance of lactic acid bacteria with the participation or not of yeasts and *Bacillus* spp. This even though lactic acid bacteria constitute a minor part of the microbiota of a vegetable or a fruit. The lactic acid bacteria often involved in lactic acid fermentation of vegetables belong to *Leuconostoc*, *Lactobacillus*, *Weissella*, *Enterococcus* and *Pediococcus* genera, being the ubiquitous and metabolic versatile *Lactobacillus plantarum* one of the most frequently isolated species. Despite the great utility of spontaneous fermentation for the conservation of raw vegetables and fruits, the chance of fermentation failure, in terms of inappropriate inhibition of spoilage or pathogen microbes, and unwanted sensory traits and nutrients composition, are significant. Therefore, contrariwise with the observed with many fermented foods, a controlled fermentation, using selected starters, is highly recommended. In this regard, the use of autochthonous cultures isolated from raw and fermented vegetables and fruits, which are adapted for the specific plant matrix, may guarantee prolonged shelf life and improved functional and sensory properties of the fermented products. Furthermore, many of these autochthonous microorganisms used as starters in plant-based fermented products are probiotics, contributing to fermented vegetables with health benefits which include the potential to modulate gut microbiota and the immune system, and to improve the functional and nutritional status through the generation of new bioactive compounds or increasing the bioavailability of existing ones.

The search for autochthonous microorganisms for the fermentation of vegetables and fruits has relatively few years of practice, starting virtually with the beginning of the 21st century. Unlike fermented foods usually produced in commercial scales, such as fermented dairy products and alcoholic beverages, for which there is an extensive knowledge concerning the used fermenting microorganisms, as well as the fermentation processes, the knowledge about fermented vegetable products, which in many regions are traditionally made, is still scarce. In this regard, the selection of the appropriate autochthonous starter microorganisms is crucial to ensure the safety and quality of the fermented vegetable foods. The selected microorganisms should overcome various challenges, such as the ability to inhibit the growth of food-borne pathogens or spoilage microorganisms, ensuring the microbiological safety and quality of fermented food, to positively modify the level or the bioavailability of nutrients and biologically active compounds and diminishing antinutrients as well, or possess probiotic potential. For the latter, besides the known beneficial properties to the host, selected microorganisms need to colonize the fermented vegetable food and dominate its indigenous microbiota under different growth and storage conditions, ensuring the recommended minimum level of viable cells at the end of product shelf-life storage (10^6-10^7 CFU/mL). The selected microorganisms also should improve the sensory characteristics of the fermented product to ensure its acceptance by consumers. Furthermore, the industry requires that the fermentation process be economically sustainable, so starter cultures must be able to ensure rapid acidification of the food matrix at low temperatures.

Fermented Leafy Vegetables, Radish, and Cucumbers

Sauerkraut (sour cabbage) is one of the most popular traditional fermented vegetables produced by fermentation of white cabbage (*Brassica oleracea* L. var. *capitata* cv. *Bronco*). It is the most used vegetable food during winter season in the Czech Republic. In sauerkraut fermentation, fresh cabbage is shredded and mixed with salt (2.3%-3.0%) and allowed for natural fermentation. The fermentation process occurs in sequential steps involving *Leuconostoc* spp. in the initial phase, and *Lactobacillus* and *Pediococcus* spp. in the subsequent phases.

Kimchi, a traditional Korean fermented food, is made from fermenting vegetables such as salted radish, cucumber, and Chinese cabbage (beachu) along with spices, including garlic, ginger, and red pepper powder. Kimchi is fermented by LAB at low temperatures ensuring proper ripening, preservation, and has strong acidic taste. The classical identification of bacterial isolates from kimchi revealed that *Lactobacillus plantarum* and *Leu. mesentroides* are the predominant LAB species.

Pao cai, one of the Chinese traditional lactic acid fermented vegetable, is usually treated in brines with high salt concentrations for long-term preservation. Four kinds of pao cai: acidified cabbage (*Brassica oleracea* L.), radish (*Raphanus sativus* L.), pak choi (*Brassica chinensis* L.), and cowpea [*Vigna unguiculata* (Linn.) Walp.] are homemade popular side dishes used in China. The dominated LAB such as *Lb. plantarum*, *Lb. pentosus, Lb. fermentum*, *Lb. brevis*, *Lb. lactis*, and *Leu. mesenteroides* are found to deplete nitrite and inhibit the growth of nitrite-reducing bacteria during pao cai fermentation. A LAB coculture (*Lb. plantarum*, *Lb.buchneri*, and *Pediococcus ethanoliduran*) fermentation in pao cai increased the content of lactic acid (sourness), sucrose, and glycine (sweetness), along with a significant quantity of gamma-aminobutyric acid.

Gundruk is a fermented leafy vegetable consumed as pickle or soup in Nepal and Himalayas regions of India. In processing of gundruk, fresh leaves of vegetables locally named as "rayo-saag" [*Brassica rapa* L. ssp. *campestris* (L.)], leaves of radish (*Raphanus sativus* L.), mustard [*Brassica juncea* (L.) Czern], cauliflower (*Brassica oleracea* L. var. *botrytis* L.), cabbages (*Brassica oleracea* L. var. *capitata* L.) are wilted for 1-2 d.

Sunki is a traditional, homemade fermented vegetables produced in the Kiso District of Japan, produced from the leaves of otaki-turnip. Otaki-turnip is boiled, mixed with zumi (a wild small apple), and dried sunki from previous year, and allowed to ferment for one or two months. It is produced in every house in Kiso during late autumn and winter. The bacterial community is found stable, and *Lb. delbrueckii*, *Lb. fermentum*, and *Lb. plantarum* are dominant LAB throughout the fermentation process. The dominant microorganisms in fermented sunki are reported to produce significant amounts of lactic acid.

Nozawana-zuke is a low-salt pickle vegetable prepared from mustard, locally called Nozawana (*Brassica campestris* var. *rapa*) and consumed by the majority of Japanese people. Generally, the LABs isolated from the spontaneous Nozawana-zuke pickle fermentation are: *Lb. plantarum*, *Lb. brevis*, *Leu. mesenteroides*, *Pediococcus pentosaceus*, and *Enterococcus faecalis*.

Tursu, a traditional fermented Turkish pickle, is made from vegetables such as cabbage, cucumber, carrot, pepper, turnip, eggplant, and beans. LAB involved in the fermentation of tursu are *Leu. mesenteroides*, *Lb. plantarum*, *Lb. brevis*, and *Pediococcus pentosaceus*.

Sinki is a nonsalted fermented radish taproot product traditionally consumed as pickle in some north-eastern states of India, Nepal, and Bhutan. The optimum fermentation time is usually 12 d at 30℃ when the pH of the fermenting mass drops from 6.7 to 3.3. The fermentation is initiated by heterofermentative *Lb. fermentum*, followed by heterofermentative *Lb. brevis*, and finally succeeded by homofermentative *Lb. plantarum*. However, the enzymatic activities of LAB such as peptidase, lipase, and esterase help in the development of flavor in fermented sinki.

Khalpi is a fermented cucumber (*Cucumis sativus* L.) eaten as pickle by adding mustard oil, salt, and powdered chillies. During the preparation, ripened cucumbers are cut into suitable pieces, sun dried for 2 d, and then put into a bamboo vessel, locally called "dhungroo" and made air tight before naturally fermenting for 4-7 d. Similar to khalpi, Jiang-guais is a popular traditional fermented cucumber in Taiwan Province.

Fermented Roots and Tuber Crop Products

Tuber crops such as sweet potato, cassava, yams, and elephant foot yams are considered as vegetables. Gari is the most popular fermented food consumed by the people of Nigeria and is derived from solid state fermentation of tapioca (syn.cassava). The initial stage of gari fermentation is dominated by *Corynebacterium* sp. with LAB succession (*Lb. acidophilus*, *Lb. casei*, *Lb. fermentum*, *Lb. pentosus*, *Lb. plantarum*, respectively).

Lafun is an African cassava fermented food product obtained by soaking peeled cassava tuber in water at ambient temperature (28-32℃) for 2-5 d. The dominated microflora such as *Lb. plantarum*, *Leu. mesenteroides* contribute cell wall-degrading enzymes (polygalacturonases and polygalacturonate lyases), and *Corynebacterium* sp. reported to contribute pectinolytic enzymes for softening cassava roots during lafun fermentation.

Fermented Vegetable Legume (Soybean) Products

Soybean, originated in Eastern Asia a 1,000 years ago, is considered as one of the important protein-rich (around 40%) vegetable legume. Additionally, fermented soybean foods contain various functional components, i.e., peptides and isoflavonoids. As reported by *Health* magazine in 2006, soybean products are classified as one of the world's top five healthiest foods. Additionally, soybean is well known to be useful for preventing obesity, diabetes, heart disease, and breast cancer.

All foods that contain proteins are subjected to conditions enabling microbial or biochemical activity; Biogenic amines (BAs) can be expected in them, including fermented and nonfermented soybean products. There are different types of soybean products (e.g., soy milk, soy curd, soy sauce, soybean paste) in which BAs can be detected. Fermentation is one of the major processes used in the production of food from soybeans. Primary sources of BAs in the soy foods include fermented foods such as soy sauce (China), miso (Japan), natto (Japan), stinky tofu (China), and tempeh (Indonesia).

Microorganisms form BAs during the fermentation process of soybean products. The microbial spoilage of food may be accompanied by the increased production of decarboxylases. Therefore, the presence of BAs might serve as a useful indicator of food spoilage in soybeans. Several species of molds, yeasts, and LAB are involved in the soy fermentation processes. Since soybeans are rich in protein, the synthesis of amines is predicted during the fermentation process.

Several studies have shown that BAs in fermented soybean are most likely formed by the fermenting lactic microflora, and histamine and tyramine were found at various concentrations. The variability of BAs levels in the fermented soy products had been attributed to the variations in production processes.

Soy paste (Doenjang) is a traditional Korean fermented soybean paste. Its name means "thick paste" in Korean language. Doenjang is produced through the fermentation of soybeans by naturally occurring bacteria, yeast, and mold, and has been consumed for centuries as a protein-rich source. This paste contains a relatively high concentration of amino acids degraded from soybeans and may be a source for BAs formation.

Chunjang is a Korean and Chinese traditional fermented sweet bean sauce produced by the fermenting steamed soybeans, flour, and salt by naturally occurring bacteria (most strains identified as *B. subtilis*).

Soy sauce is a condiment produced from paste of boiled soybeans, roasted grains, brine, and fermented by *Aspergillus oryzae* or *Aspergillus sojae* molds. Soy sauce is a traditional component in East and Southeast Asian cuisines, where it is used as a condiment and in cooking. The flavor in soy sauce may develop gradually during the fermentation and aging.

Miso is a traditional Japanese paste. It is produced by fermenting rice, barley, and soybeans with salt and the fungus. This paste is used for sauces and spreads, pickling vegetables or meats and Japanese culinary staple. In miso, tyrosine decarboxylase bacteria have been identified as *Enterococcus faecium*, *Lb. bulgaricus*, and histamine decarboxylase have been associated with *Lactobacillus* sp. and *Lactobacillus sanfrancisco*.

Natto is a traditional Japanese soy product made from soybeans fermented with *B. subtilis*. Researchers identified some histamine-producing bacteria belonging to *Lactobacillus* species in natto products manufactured in Taiwan. However, *Bacillus subtilis* is highly capable of producing β-phenylethylamine (30 mg/kg) and tyramine (100 mg/kg) in natto products.

Sufu (furu) is a kind of Chinese traditional fermented soybean curd made from soybean that is easily digested and a nutritious protein. The pure starter culture consists mainly of molds (*Actinomucor*, *Mucor*, and *Rhizopus*) or bacteria (*Micrococcus* and *Bacillus* spp.) responsible for sufu fermentation. The process of sufu manufacture is itself carried out under nonsterile conditions, which caused microbial contamination and led to the formation of BAs. However, *Clostridium perfringens* grows in protein-rich media. Accordingly, this bacterium is often detected in the amino acid rich environment, including sufu. The type of dressing (salting and ripening) used determines the color, and subsequently the category of sufu (white, gray, brown, sesame oil, hot, and alcoholic sufu). In sufu, the commonly found BAs are putrescine, cadaverine, and tryptamine. However, the BAs (β-phenylethylamine, histamine, and tyramine) content of sufu exceeded recommended toxicity limits by a factor of approximately 0-0.17 mg/kg.

Fermented Juice

Juice is defined as "the extractable fluid contents of tissues or cells". Each juice has particular chemical, nutritional, and sensorial characteristics, depending upon the type of fruit or vegetable used. Researchers have indicated that consumption of fruits and vegetables or drinking their juices is correlated with the reduction of chronic diseases risk. Vegetable and fruit juices are sources of many bioactive phytochemicals such as vitamins, minerals, and phenolic compounds. Many studies showed that there is a correlation between the total phenols and their antioxidant activities.

The consumers' requirements in the food field changed considerably. Consumers show an increased request for foods and drinks that can stimulate well-being, which encouraged functional food production. Well-known examples of these products are the fruit and vegetable juices with probiotics. Furthermore, these beverages are also good for people with lactose intolerance.

The fermentation is considered as a low-cost process, which preserves the food and improves its nutritional and sensory characteristics. Many cultures were used as starter cultures for fermented juices and have been recognized as probiotics. Moreover, these fermentations allow introducing new products to the market.

(1) Composition and Nutritive Value of Fruit and Vegetable Juice

Fruits and vegetables represent one of the essential elements for a balanced food and are known for their promoting human health. They are often regarded as "functional food" thanks to their content rich in various micronutriments such as phenolic compounds (recognized, in particular, for their antioxidant capacity), minerals, and vitamins. Many epidemiologic studies showed a relationship between the consumption of fruits and/or vegetables and prevention of various diseases like cancer, neurodegenerative diseases, obesity, and diabetes.

The fruits' composition is similar to that of vegetables. They are characterized by very important water content (90% on average). However, carbohydrate content in fruits is higher than that in vegetables. Organic acids are the second most abundant soluble solids component of the juices, and they have an impact on their sensory properties. Malic and citric acids are the main acids. Tartaric, glutamic, oxalic, and quinic acids are also present in large quantities in many fruits. Moreover, juices represent an important mineral source (such as sodium, potassium, iron, zinc, magnesium, phosphorus, copper, calcium, and manganese), contributing to the acid-basic equilibrium in the blood and facilitating the neutralization of noxious uric acid reactions. The main juice phytochemicals are vitamins, phenolics, and pigments, which promote the chemopreventive potential of these juices. The main subtypes of polyphenols are flavonoids, tannins, and phenolic acids. Flavonoid can be divided into many classes such as flavones, flavonols, isoflavones, flavanones, and anthocyanidins. They have powerful antioxidant activities and exhibit protective effects against many chronic diseases and infectious.

(2) Juice Fermentation by Lactic Acid Bacteria

Lactic acid bacteria are extensively used in food fermentation. Their use improves the organoleptic characteristics and nutritional values of fermented products. During the fermentation, LAB transform indigestible substances into others easier to digest and produce different antimicrobial compounds. Some of these bacteria are called "probiotic" and are known to have health-promoting attributes.

Among all LAB, *Lactobacillus plantarum* is the most used species for vegetable fermentation. LAB use carbon sources and free amino acids present in the medium to produce metabolites of interest. Lactic acid is the major organic acid produced by LAB, and it has a role in the reduction of the proinflammatory cytokine secretion of Toll-like receptor (TLR) activated, bone marrow-derived macrophages and dendritic cells in a dose-dependent manner. Besides organic acids'production, many secondary metabolites are synthesized such as bioactive peptides, fatty acids, exopolysaccharides, and vitamins. LAB increase also the antioxidant activity of fermented products thanks to enzymes' activity such as β-glucosidase and esterase. Moreover, the produced phenol derivatives are usually a source of the end products' aroma.

New Words

autochthonous [ɔː'tɒkθənəs]= autochthonal; autochthonic *adj.* 土著的；土生土长的；原地形成的
pathology [pə'θɒlədʒɪ] *n.* 病理学；病状
exopolysaccharide [eksə'pɒlɪsækærɪd] *n.* 胞外多糖
mustard ['mʌstəd] *n.* 芥末酱；芥末黄，深黄色；芥菜
potherb ['pɒtˌhɜːb] *n.* 野菜；调味香草
phenolic [fɪ'nɒlɪk] *adj.* 酚的；酚醛树脂的
n. 酚醛树脂
flavonoid ['fleɪvənɔɪd] *n.* 黄酮类；[有化]类黄酮
phytoestrogen [ˌfeɪtəʊ'estrədʒən] *n.* 植物雌激素
inflammatory [ɪn'flæmətrɪ] *adj.* 炎症性的；煽动性的；激动的
brine [braɪn] *n.* 卤水；盐水
vt. 用浓盐水处理（或浸泡）
glycine ['glaɪsːn] *n.* 甘氨酸
lipase ['laɪpeɪz] *n.* 脂肪酶；脂肪分解酵素
esterase ['estəreɪs] *n.* 酯酶；甘油松香酯
tapioca [ˌtæpɪ'əʊkə] *n.* 木薯淀粉
β-phenylethylamine ['fiːnɪlˌiːθaɪlə'miːn] *n.* β-苯（基）乙胺
beverage ['bevərɪdʒ] *n.* 饮料
neurodegenerative [ˌnjʊːrəʊdɪ'dʒenərətɪv] *adj.* 神经变性的

Notes

1) The elaboration of fermented plant-based foods and beverages constitute an alternative

for its sustainable use, also transforming them in carriers for delivery of potential probiotics of value to consumers suffering from allergy to milk proteins or lactose intolerance.

参考译文：为了使新鲜果蔬可持续利用，植物性发酵食品和发酵饮料的精制可作为替代品，还可以将其作为载体，为患有乳蛋白过敏或乳糖不耐受的消费者提供有价值的潜在益生菌。

2) Fermentation of fruits and vegetables, which is the main and simple alternative to heat or chemicals for guarantee the safety of vegetable products, may also lead to positive changes in the concentration or composition of vitamins, amino acids, bioactive peptides, or phytochemicals.

参考译文：果蔬的发酵是一种可替代加热或化学物质的主要且简单的加工方式，能够保证产品的安全性，也可能促使果蔬中的维生素、氨基酸、生物活性肽或植物化学物的浓度或组成发生积极变化。

3) This microbiota, composed mainly by beneficial microorganisms, including yeasts (*Saccharomyces*, *Pichia*, *Candida*, *Torulaspora genera*), fungi (*Rhizopus* spp.), and aerobic (*Bacillus* spp. and *Acetobacter* spp.) and anaerobic (lactic acid bacteria) bacteria, is usually responsible for the spontaneous fermentation of raw vegetables and fruits, contributing to their preservation and stability.

参考译文：这种微生物群主要由有益微生物组成，包括酵母菌（酵母属、毕赤酵母属、念珠菌属、孢圆酵母属）、真菌（根霉属）、好氧细菌（芽孢杆菌属和醋酸杆菌属）和厌氧细菌（乳酸杆菌）。这些微生物通常负责蔬菜和水果的自然发酵，有助于果蔬的储存和稳定性。

4) Therefore, contrariwise with the observed with many fermented foods, a controlled fermentation, using selected starters, is highly recommended.

参考译文：因此，与观察到的许多发酵食品相反，强烈推荐使用精选的发酵剂进行可控发酵。

5) Fermentation is one of the major processes used in the production of food from soybeans. Primary sources of BAs in the soy foods include fermented foods such as soy sauce (China), miso (Japan), natto (Japan), stinky tofu (China), and tempeh (Indonesia).

参考译文：发酵是生产大豆制品的主要工艺流程之一。大豆食品中生物胺的主要来源包括发酵食品，如酱油（中国）、味噌（日本）、纳豆（日本）、臭豆腐（中国）和豆鼓（印度尼西亚）。

6) Moreover, juices represent an important mineral source (such as sodium, potassium, iron, zinc, magnesium, phosphorus, copper, calcium, and manganese), contributing to the acid-basic equilibrium in the blood and facilitating the neutralization of noxious uric acid reactions.

参考译文：此外，果汁是一种重要的矿物质来源（如钠、钾、铁、锌、镁、磷、铜、钙和锰），有助于血液中的酸碱平衡，并促进中和有害尿酸的反应。

Reading Material 1 Vinegar

In the food industry, vinegar is used mainly as an acidulent (*n.* 酸化剂), a flavoring agent, and a preservative, but it also has many other food processing applications. It is found in hundreds of different processed foods, including salad dressing, bread and bakery products, pickled foods, canned foods, and sauces.

Vinegar must be made from one of various types of ethanol-containing solution. The most common starting materials are grape and rice wine, fermented grain or malt, and fermented apple cider. Distilled ethanol is also permitted as a substrate for vinegar manufacture. Importantly, vinegar must, by definition, result from the acetous fermentation of ethanol, and contain at least 4% acetic acid. Usually, the ethanol concentration is less than 0.5%, and the pH is between 2.0 and 3.5.

The raw material determines the name of the vinegar (e.g., red wine vinegar, apple cider vinegar, malt vinegar). However, the starting material also has a profound influence on the flavor and overall quality attributes of the vinegar. Although the predominant flavor of all vinegars is due to acetic acid, other flavors, specific to the ethanol source or the means of its manufacture, may also be present. In addition, some vinegar may also contain herbs, added before or after the fermentation.

The manufacture of vinegar consists of two distinct processes. The first step is an ethanolic fermentation preformed mostly by yeast. In the other step, an acetogenic fermentation is carried out by acetic acid bacteria. The metabolic process involves conversion by alcohol dehydrogenase (*n.* 脱氢酶) of ethanol to acetaldehyde and conversion by acetaldehyde dehydrogenase of hydrated acetaldehyde to acetic acid.

$$C_2H_5OH \rightarrow CH_3CHO+H_2O \longleftrightarrow CH_2CH(OH)_2 \rightarrow CH_3COOH+H_2O$$

Whereas the ethanol fermentation is anaerobic, the latter is conducted under highly aerobic conditions. In fact, many of the technological advances in the vinegar fermentation have focused on ways to introduce more air or oxygen into the fermentation system.

Acetic acid bacteria, which oxidize ethanol to acetic acid and can exist at low pH values, come from the closely-related genera *Acetobacter* (*n.* 醋杆菌属) and *Gluconobacter* (*n.* 葡糖杆菌属). Pure cultures of these organisms are characterized by their high degree of variability and in industrial fermentations mixed cultures will consequently develop from a pure culture. Industrial cultures are selected to tolerate high acidity and to yield high acetate production rates. These bacteria are extremely sensitive, are killed by lack of oxygen and lack of ethanol and are also damaged by acetate and ethanol concentration gradients. Sensitivity to lack of oxygen increases with increasing total concentration of acetic plus ethanol. Nevertheless, with efficient aeration, an oxygen utilization of 80% can be achieved without adverse effects on the fermentation. Over-oxidation, that is conversion of acetic acid to CO_2 and H_2O, can be avoided by maintaining acetic acid concentrations above 6% and avoiding total depletion of ethanol.

The conversion step of ethanol substrate to acetic acid can be performed by one of several methods. These processes are generally referred to as the open vat method, the trickling generator process, and the submerged fermentation process. The latter two are now the more widely used, especially for large vinegar manufacturers, since they can perform in continuous mode and can reduce the fermentation time from several weeks to a matter of days.

The earliest example of trickling generator used on an industrial scale was in 1823. In this system, which is still widely used, ethanolic substrates are circulated or trickled through cylindrical fermentation vessels or vats containing inert packing materials (*n.* 惰性填充物), such as curled wood shaving， wood staves, or corn cobs (Figure 3.1). These materials were used not to contribute flavor but provide a large surface area. As the inoculated substrate or feedstock passes from the top to the bottom of the vessel, the total surface area of the liquid material increases as it moves around and between the particulate packing materials. Growth of acetic acid bacteria will then occur at the air-liquid interface, such that the ethanol concentration decreases, and the acetic acid concentration increases during the transit of substrate from top to bottom. Holes can be drilled (*v.* 钻孔) into the side of the vessel to ensure that aeration is adequate. When the substrate reaches the bottom section, it may be returned to the top until the effluent is sufficiently acidic to be called vinegar. Alternatively, a second tank can be used to increase the oxidation rate and product throughput. The inoculated mash is fermented first in tank 1, then passed, with aeration, into tank 2, and back and forth until all of the ethanol is converted to vinegar. About 3 d are required to convert a 12% (*V/V*) ethanol solution to vinegar containing 10% to 12% acetic acid.

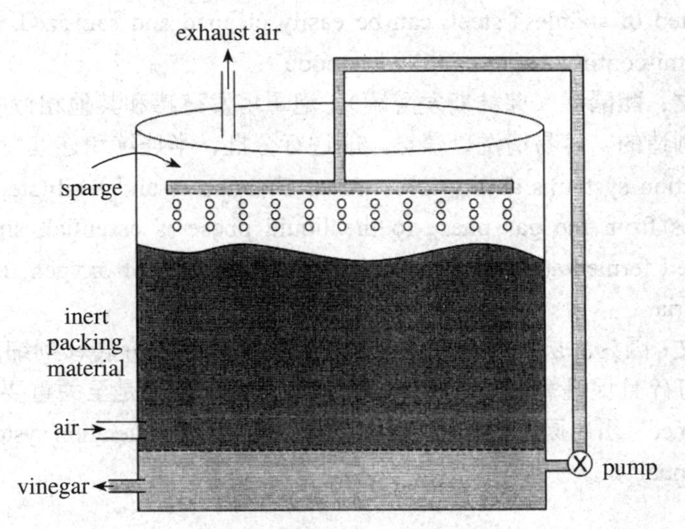

Figure 3.1 Trickling generator system

Advances in biotechnology have led to the development of modern industrial fermentors capable of rapid, high throughput bioconversion processes. Although submerged (*adj.* 在水中的，淹没的) fermentation systems are now widely used for many industrial fermentation processes, they were actually developed by the vinegar industry more than fifty years ago. The acetator (the

Frings fermentor), bubble fermentor, and other similar units are constructed of stainless steel, can be easily cleaned and sanitized, and can operate in batch, semi-continuous, or continuous mode. The most important feature of the submerged fermentation systems is their ability to provide rapid and efficient aeration. For example, the Frings Acetator is equipped with turbines (*n.* 涡轮) that mix the liquid with air or oxygen and deliver the acerated mixture at very high rates inside the fermentor. The aeration system's ability to break up air bubbles and facilitate transfer of oxygen molecules from the gas phase to the liquid phase is essential, since the success of submerged fermentation systems relies on the transfer of oxygen from the medium to the bacteria. In general, the starting material contains both ethanol and acetic acid. After a single cycle of 16-24 h, the ethanol concentration will be reduced from 5% to less than 0.5%, and acetic acid will increase from around 7% to 12%. Up to half of the vinegar is then removed and replaced with fresh ethanol feedstock.

Notes

1) These bacteria are extremely sensitive, are killed by lack of oxygen and lack of ethanol and are also damaged by acetate and ethanol concentration gradients.

   参考译文：这些细菌非常敏感，缺乏氧气和乙醇时会死亡，也会被乙酸和乙醇浓度梯度损伤。

   句中“are killed”和“are also damaged”是并列谓语。“lack of”缺乏。

2) The acetator (the Frings fermentor), bubble fermentor, and other similar units are constructed of stainless steel, can be easily cleaned and sanitized, and can operate in batch, semi-continuous, or continuous mode.

   参考译文：酿醋罐（弗林斯发酵罐）、泡罩塔发酵罐和其他相似的发酵设备都是由不锈钢制造的，容易清洗、消毒，能进行分批、半连续和连续发酵。

3) The aeration system's ability to break up air bubbles and facilitate transfer of oxygen molecules from the gas phase to the liquid phase is essential, since the success of submerged fermentation systems relies on the transfer of oxygen from the medium to the bacteria.

   参考译文：因为深层发酵系统的成功依赖于氧气从媒介转移到细菌，故通风系统将空气泡打碎且促进氧气分子从气相转移到液相的能力是至关重要的。

   句中“since”引出原因状语从句。主句中主语是“The aeration system's ability”，表语是“essential”。

Exercise

1. True or false.

1) Vinegar must, by definition, result from the acetous fermentation of ethanol, and contain

at least 4% acetic acid. ( )

2) The raw material only determines the name of the vinegar, has not a profound influence on the flavor and overall quality attributes of the vinegar. ( )

3) The ethanol fermentation is anaerobic, whereas acetic acid fermentation is conducted under highly aerobic conditions. ( )

4) Over-oxidation, that is conversion of acetic acid to CO_2 and H_2O, cannot be avoided by maintaining acetic acid concentrations above 6% and avoiding total depletion of ethanol. ( )

5) The inert packing materials, such as wood staves or corn cobs, can be used not to provide a large surface area, but contribute flavor of vinegar. ( )

2. Translate the following sentences into Chinese.

1) Vinegar must be made from one of various types of ethanol-containing solution. The most common starting materials are grape and rice wine, fermented grain or malt, and fermented apple cider. Distilled ethanol is also permitted as a substrate for vinegar manufacture.

2) Although the predominant flavor of all vinegars is due to acetic acid, other flavors, specific to the ethanol source or the means of its manufacture, may also be present.

3) Industrial cultures are selected to tolerate high acidity and to yield high acetate production rates.

4) In general, the starting material contains both ethanol and acetic acid. After a single cycle of 16-24 h, the ethanol concentration will be reduced from 5% to less than 0.5%, and acetic acid will increase from around 7% to 12%.

Reading Material 2 Fermented Soy Bean Foods

Fermented soybean foods made with *Bacillus subtilis* cells are produced in China, and they are called *dou chi*. They include salted, sweet, and nonsalted types. The salted type of *dou chi* (*xian-dou chi*) contains 10% to 20% salt to inhibit their putrefaction by contaminating bacteria. The most typical sweet *dou chi* is used as a seasoning for *Beijing duck*. Nonsalted *dou chi* has been developed into various kinds of natto. Food of this type is called *itohiki-natto* (hereafter shortened to "*natto*") in Japan, *kinemain Nepal* and *Myanmar*, *tuanao* in Thailand, and *chungkuk-jang* in Korea. *Natto* is produced only with *B. subtilis* (*natto*).

These fermented soybeans are consumed in a variety of forms. For instance, *tuanao* is used as a raw ingredient in salads. *Chungkuk-jang*, which contains cayenne peppers and garlic, is used as an ingredient for a Korean soup called *chige*. In Japan, *natto* is mixed with soy sauce, sliced Welsh onion (similar to stone leeks), mustard, dried seaweed, and/or raw egg. This seasoned *natto* is usually eaten with rice in Japan. A looser *natto* (*hikiwari-natto*) and a dried type of *natto* are also used. Before World War Ⅱ, the transportation infrastructure was not well developed in

Japan. Hence, dried, hard *natto*, having a long shelf life, was usually manufactured and consumed. Of the three types of *natto*, *itohiki-natto* is the most popular at present in Japan. *Hikiwari-natto* is used for preparing sushi. Consumption of *natto* increased compared to other soybean foods during the 1990s, generating sales of 160 billion yen in 1996.

B. subtilis, B. subtilis(natto), and *natto* are considered to have potential as probiotics. Ingestion of bacterial cells probably affects the intestinal microflora and the mucosal immune system. *B. subtilis (natto)* cells produce many enzymes and vitamin K_2 (menaquinone-7). A serine protease, subtilisin, can degrade soybean allergens and shows fibrinolytic activity. Ingestion of vitamin K_2 will help coagulant activity and prevent osteoporosis. *Natto* contains the phytoestrogens (isoflavones) (*n.* 异黄酮) that originate in soybeans. Isoflavones seem to have preventive effects on breast and prostate cancer, osteoporosis, menopausal (*adj.* 绝经期的；更年期的) symptoms, and heart disease.

Lactobacillus spp. and *Bifidobacterium* spp. are mainly used as probiotics for humans and animals. However, other bacteria and fungi can also be used as probiotics. For example, *Bacillus* spp., *Enterococcus* spp., *Streptococcus* spp., and *Saccharomyces cerevisiae* seem to have potential as probiotics. In the screening and selection of certain microbial strains as probiotics, phenotype (*n.* 显型；表现型) and genotype (*n.* 基因型；遗传型) stability, carbohydrate and protein utilization patterns, safety, acid and bile stability, adhesion characterization, production of antimicrobial substances, antibiotic resistance patterns, immunogenicity, and viability and properties during processing and storage are considered to be important. *B. subtilis* spores are relatively resistant to oxygen, active oxygen species, acid, drying, and heating compared to other bacteria. *B. subtilis* can also grow under O_2-reduced conditions. These characteristics are desirable for potential probiotics. Unfortunately, however, *B. subtilisis* not strongly resistant to bile acid, and is not a predominant bacterium in the human intestine. Several reports have demonstrated the effects of orally administered *B. subtilis* on the intestinal microflora, body weight gain, and increased feed efficiency of animals and birds. These results indicate that ingestion of live *B. subtilis* cells can actually improve the intestinal microflora. When weanling piglets were fed a diet including spores of *B. subtilis* (*natto*), the changes in intestinal microflora varied depending upon the region of the intestine examined. In the jejunum, the numbers of *Streptococcus* spp. and *Bifidobacterium* spp. increased, whereas no differences were observed in the colon, when compared with the control diet group. When turkeys were fed *B. subtilis* culture, body weight gain and cumulative feed efficiency significantly increased, both by 2.5%. When chickens were given *B. subtilis*, the detection rate of the intestinal pathogen *Campylobacter jejuni* decreased in the laboratory portion of the experiment. The cell number of *Salmonella typhilimurium* also decreased. In a field trial, feeding a *B. subtilis* strain decreased the cell number and detection rates of intestinal *Enterobacteriaceae*, *Clostridium perfringens*, and *Campylobactersp*. When sows and gilts were fed an experimental diet containing *B. subtilis*, the number and detection rates of fecal *Bifidobacterium* spp. and *Lactobacillu* spp. increased, but

Streptococcus spp., *Enterobacteriaceae*, *Clostridium perfringens*, and *Bacteroidaceae* decreased. The diarrhea rate of the piglets up to 10 d old and mortality rate up to 25 d old also decreased. When mice were intubated with intact and autoclaved *B. subtilis* (*natto*) spores for 8 d, only intact spores changed the fecal microflora, and the patterns of the changes differed depending upon the diets fed. Feeding a diet including egg white decreased fecal *Lactobacillus* spp., although the administration of *B. subtilis* (*natto*) spores inhibited the decrease. On the other hand, feeding a diet including casein and administering *B. subtilis* (*natto*) spores increased only *Bacteroidaceae* but not *Lactobacilli*.

Ingestion of soybean food *natto* (50 g *natto*/d) significantly affected the composition and metabolic activity of the human fecal microflora. Ingestion of *natto* increased the number of *B. subtilis* (*natto*) and *Bifidobacterium* spp. (the latter increased from 15% of the total bacterial count before consumption to 39% after 14 d consumption), although it decreased the number and detection rates of lecithinase positive *clostridia* including *C. perfringens*. The concentrations of fecal acetic acid, total organic acids, and succinic acid increased, while fecal concentrations of indole, ethylphenol, and skatol decreased. Fecal ammonia, cresol, and fecal pH values also decreased.

Mechanisms of the above effects have not been clarified. However, germination or some metabolites from *B. subtilis* cells seem to be necessary to explain their effects, because it was shown that in mice, administration of autoclaved spores did not affect the fecal (*adj.* 排泄物的；残渣的；糟粕的) microflora. The possibility of germination of *Bacillus* spp. spores in the intestine has been examined. When *B. subtilis* (*natto*) spores were inoculated in the ligated loops of the ileum of dogs, some spores did germinate, but died after germination (*n.* 发芽；萌芽；共生；晶核化).

It has been shown that *B. thuringiensis* spores germinate in the gut fluid of the tobacco horn worm. *B. subtilis* spores also germinate in the mice gut. In this report, the number of spores excreted in the feces of the mice was, in some experiments, larger than the number of spores inoculated. However, this is inconsistent with the report that vegetative cells of *B. subtilis* cells could not be detected when spores were inoculated in mice that were left without food for 16 h.

Live *Bacillus* cells could be detected only in some organs after ingestion (*n.* 摄取；吸收；咽下). In general, when foods are ingested, the pH value in the stomach sometimes increases to 3-4. Spores of *Bacillus* spp. appear to be resistant to such pH values. Some of the *B. subtilis* spores ingested together with other food may be able to sustain their viability and germinate in the upper intestine once the surrounding pH value is neutralized. This then allows them to produce probiotic activity. Catalase and subtilisin have been proposed as the active molecules responsible for the effects of *B. subtilis* (*natto*) on intestinal (*adj.* 肠的) microflora. The growth of three strains of *Lactobacilli* co-cultured aerobically with *B. subtilis* (*natto*) has been examined. Addition of *B. subtilis* (*natto*) to the culture medium in vitro resulted in an increase in the number of viable cells of all *Lactobacilli* tested. Both catalase and *B. subtilis* (*natto*) enhanced the growth of *Lb. reuteri*, whereas *B. subtilis* (*natto*), but not catalase, enhanced the growth of *Lb. acidophilus*. In a medium containing 0.1 mmol/L hydrogen peroxide, its toxic effect on *Lb. reuteri* was abolished by catalase or *B. subtilis* (*natto*). Catalase has been reported to exhibit a growth-promoting effect on *Lactobacilli*. The viability of *Lactobacilli* readily decreases in the presence of active oxygen species. The decrease of viable cell number is partly attributable to the fact that *Lactobacilli* do not generally produce a defense molecule against active oxygen species. However, aerobic bacteria, including *B. subtilis* (*natto*), can produce catalase. Vegetative cells of *B. subtilis* primarily produce catalase-1 in the logarithmic phase of growth, and additionally produce catalase-2 and catalase-3 as growth progresses. Intact *B. subtilis* spores contain only catalase-2 in the spore (*n.* 孢子；*v.* 产生孢子) coat. Some other anaerobic bacteria in the intestine, such as *Escherichia coli*, *Bacteroides* spp., and *Eubacterium* spp., also produce catalase.

It may be important for these bacteria to scavenge hydrogen peroxide to colonize the intestine where active oxygen species are produced. The addition of a serine protease, subtilisin, from *B. licheniformis* to culture medium improved the growth and viability of *Lb. reuteri* and *Lb. acidophilus* in the absence of hydrogen peroxide.

B. subtilis (*natto*) secretes two serine proteases, subtilisin NAT with an isoelectric point (pI) of 8.7 and a 90 ku serine proteinase (*n.* 蛋白酶) (pI 3.9). Taken together, these results indicate that *B. subtilis* (*natto*) can enhance the growth and viability of *Lactobacilli* possibly through production of catalase (*n.* 接触酵素；过氧化氢酶) and subtilisin.

Natto is a simple, low-priced, popular soybean food made by fermenting cooked soybeans with *B. subtilis* (*natto*) in Japan. *Natto* has characteristic aroma and stickiness.

The consumption of *natto* products has increased during the 1990s in Japan. *Natto* contains many nutrients originating from soybeans and the metabolites of *B. subtilis* (*natto*) cells, which show many physiological functions. In 1999, the FDA in the US approved the use of health claims for soy protein related to the reduction of the risk of coronary heart disease by lowering blood cholesterol levels when 25 g of soy protein per day are consumed. In addition, many physiological functions of *B. subtilis* cells have recently been reported.

Although *B. subtilis* (*natto*) is not a predominat bacterium in the human intestine, it is thought to have the potential as a probiotic. Some pharmaceutical products containing *B. subtilis* spores have been sold and utilized in Japan and European countries. Both scientific researchers and consumers are paying attention now to *natto* and *B. subtilis* cells. However, their effects and mechanisms are not clearly understood yet. Further careful studies are necessary to obtain data on their effects, safety, and efficacy. We hope that the studies on the nutrition and physiological function of *natto* products, *B. subtilis* cells, and related products will progress steadily and that such products will become popular all over the world.

Notes

1) However, this is inconsistent with the report that vegetative cells of *B. subtilis* cells could not be detected when spores were inoculated in mice that were left without food for 16 h.

参考译文：但是这与报道不一致，报告记载当枯草芽孢杆菌的孢子接种空腹16小时的小鼠体内后，检测不到其生长细胞。

2) We hope that the studies on the nutrition and physiological function of *natto* products, *B. subtilis* cells, and related products will progress steadily and that such products will become popular all over the world.

参考译文：我们希望关于纳豆、枯草芽孢杆菌及其相关产品的营养与生理学功能的研究将稳步前进，希望这些产品能在全世界受到欢迎。

Exercise

1. Translate the following sentences into Chinese.

1) Fermented soybean foods made with *B. subtilis* cells are produced in China, and they are called *dou chi* (or *dauchi*). They include salted, sweet, and nonsalted types.

2) It may be important for these bacteria to scavenge hydrogen peroxide to colonize the intestine where active oxygen species are produced.

2. True or false.

1) *Lactobacillus* spp. and *Bifidobacterium* spp. are the only stains used as probiotics for humans and animals. (　　)

2) Ingestion of bacterial cells don't affects the intestinal microflora and the mucosal immune system. (　　)

Reading Material 3　Probiotics

Probiotics (*n.* 益生菌) are described as "live microorganisms which, when administered in adequate amounts, confer a health benefit on the host". Examples of health benefits associated with the consumption of probiotics include a decrease in rotavirus (*n.* 轮状病毒) shedding in infants, reductions in antibiotic-associated diarrhea (*n.* 腹泻), reduction in the incidence of childhood atopic eczema (*n.* 异位性湿疹), and management of inflammatory (*adj.* 炎症性的) bowel diseases. Foods containing probiotics, such as fermented milks, yogurts, and cheese, fall within the functional food category, which includes any fresh or processed food claimed to have health-promoting and/or disease-preventing properties beyond the basic nutritional function of supplying nutrients. The area of probiotics, prebiotics (*n.* 益生元), and synbiotics (*n.* 合生元) represent the largest segment of the functional food market in Europe, Japan, and Australia.

In order to exert health benefits on the host, probiotics must be able to grow in the human intestine (*n.* 肠道), and therefore, should possess the capability to survive passage through the GIT, which involves exposure to hydrochloric acid (*n.* 盐酸) in the stomach and bile (*n.* 胆汁) in the small intestine. *Lactobacillus* and *Bifidobacterium* species are ideal probiotic candidates for incorporation into foods for human consumption. These microorganisms are known inhabitants of the GIT and share a number of common traits such as acid and bile tolerance, the ability to adhere to intestinal cells, and GRAS (generally regarded as safe) status. A major challenge associated with the application of probiotic cultures in the development of functional foods is the retention of viability during processing. Given that probiotics are generally of intestinal origin, many such strains of bacteria are unsuitable for growth in dairy-based media, and are inactivated upon exposure to high temperatures, acid, or oxygen during dairy and food processing. The survival of Bifidobacteria (*n.* 双歧杆菌) during processing can be particularly challenging, as these are strictly anaerobic microorganisms with complex nutritional requirements. Maintaining the viability minimum numbers of probiotic cultures present in the final product recommended to be 10^7 CFU/mL or even higher and the activity of probiotic cultures in foods to the end of shelf life are two important criteria that must be fulfilled in order to provide efficacious probiotic food products.

Fermented dairy foods, including milk and yogurt, are among the most accepted food carriers for delivery of viable probiotic cultures to the human GIT. Because high levels of probiotics are recommended for efficacy of these products, preparation of bulk cultures is required. However, because probiotics are normally of intestinal origin, these cultures exhibit poor growth rates in synthetic and milk-based media. Spray drying and freeze drying are useful means of introducing the probiotic culture into these food systems. The use of such approaches in preparing cultures may impair viability and probiotic functionality due to the extent of cell injury that may occur during these processes upon exposure to extreme heating and drying or freezing and drying.

Recently, probiotics have been defined as "living microorganisms that resist gastric (*n.* 胃液), bile, and pancreatic (*adj.* 胰腺的) secretions; attach to epithelial (*adj.* 上皮的) cells; and colonize the human intestine". The definition of probiotics has changed from the original one of being a live active culture beneficially affecting the host by improving its intestinal microbial balance, to the current concept based on the specific effects of clearly defined strains. This focuses attention on demonstrated clinical (*adj.* 临床的) effects, which may be mediated either through probiotic effects on the intestinal immune system or through modulation of the gut microbiota at specific locations.

The microbes administered should be safe, have GRAS status, and also have a long history of safe use in foods. All probiotic strains should have nonpathogenic (*adj.* 非致病的) properties, and ideally should exhibit tolerance to antimicrobial substances, but should not be able to transmit such resistance to other bacteria.

Adherent probiotic strains are desirable because they have a greater chance of becoming established in the GIT, thus enhancing their probiotic effect. Adhesion to the intestinal mucosa is considered important for immune (*adj.* 免疫的) modulation (the intestine being the largest immune organ of the body), and for pathogen exclusion by stimulating their removal from the infected intestinal tract.

The mechanisms by which functional microbes and ingredients affect human gut (*n.* 肠道,

内脏) health are still largely unknown. The knowledge acquired by genomics on the genetics and physiology of a probiotic strain can be used for strain improvement. For example, it may be possible to eliminate an undesirable trait from a strain by simple mutagenesis (*n.* 诱变). Alternatively, the use of recombinant DNA technology may allow the production of strains with exactly the correct combination of properties. Such improved understanding may offer solutions to many of the challenges that currently hamper the commercial development of functional foods containing particular probiotic strains by allowing the preconditioning of strains to be used for industrial fermentation.

Notes

1) Spray drying and freeze drying are useful means of introducing the probiotic culture into these food systems. The use of such approaches in preparing cultures may impair viability and probiotic functionality due to the extent of cell injury that may occur during these processes upon exposure to extreme heating and drying or freezing and drying.

   参考译文：喷雾干燥和冷冻干燥是将益生菌菌株引入这些食品体系的有效方法。由于经历极端加热干燥或冷冻干燥过程中可能造成一定程度的细胞损伤，在使用此类方法制备菌株时可能会损害益生菌活性和功能。

2) Adherent probiotic strains are desirable because they have a greater chance of becoming established in the GIT, thus enhancing their probiotic effect.

   参考译文：黏附益生菌菌株是理想的，因为它们在胃肠道中建立的机会更大，从而增强其益生菌效果。

Exercise

1. Translate the following words into English.

益生菌；厌氧微生物；大肠埃希菌；免疫器官；功能性食品

2. Translate the following sentences into Chinese.

1) Probiotics are described as “live microorganisms which, when administered in adequate amounts, confer a health benefit on the host”.

2) Recently, probiotics have been defined as “living microorganisms that resist gastric, bile, and pancreatic secretions; attach to epithelial cells; and colonize the human intestine”.

3. Translate the following sentences into English.

1）许多不同属的微生物用作益生菌，包括乳杆菌属、双歧杆菌属、丙酸杆菌、芽孢杆菌属、埃希菌属、肠球菌属和酵母属。

2）这个定义强调，益生菌可以是无活力的细胞或细胞的一部分，因为在这些形式与某些发酵终产物和酶一样，已经显示出产生健康益处。

Chapter 4

Food Additives and Enzymes

Lesson 1 Food Additives

Food additives are substances added to food to preserve flavor or enhance its taste, appearance, or other qualities. The proper use of food additives plays an important role in modern food industry. Firstly, food additives not only function in food preservation, but also help to enhance certain qualities of food such as color, flavor and flexibility, as well as improving the nutrition of food. Secondly, the use of food additives such as defoaming agents, stabilizers and coagulants in food is conducive to the operations during food processing. Additionally, food additives benefit the commercial convenience by extending the shelf life and working in the manufacturing process, through packaging, or during storage or transport.

Definitions of food additives (direct food additives, ingredients added to foods for a specific purpose) vary among government agencies and organizations. World Health Organization (WHO): food additives means any substance not normally consumed as a food by itself and not normally used as a typical ingredient of the food whether or not it has nutritive value, the intentional addition of which to food for a technological (including organoleptic) purpose in the manufacture, processing, preparation, treatment, packing, packaging, transport, or holding of such food results or may be reasonably expected to result (directly or indirectly), in it or its by-products becoming a component of or otherwise affecting the characteristics of such foods.

Till now, there are more than 25,000 compounds of food additives being used all around the world. According to the compositions, food additives are generally divided into two major categories of natural additives and synthetic additives. Wherein, natural food additives are mainly produced by purifying the ingredients from plant or animal sources. While chemically synthesized additives are based on chemical raw materials, from which organic or inorganic matter can be extracted and purified. Additionally, based on their functions, food additives can be divided into several groups such as colorants, preservatives, antioxidants, sweeteners, thickeners and so on (Table 4.1).

Table 4.1 Typical representatives of food additives

Category		Common substance	Mainly used
Colorants	Natural	Natural carotene	Dairy products, frozen drinks, processed fruits, dried vegetables and soybean products
		Anthocyanins	Jams, sugar confectioneries, jellies, soft drinks and frozen products
		Curcumin	Beverages, sauces and confectionery
		Canthaxanthin	Tanning pills, fruit-spreads, candies, sirups, sauces, carbonated drinks
	Artificial	Tartrazine	Soups, sauces, ice creams, ice lollies, sweets, chewing gum, marzipan, jam, jelly, marmalade and mustard yogurt
		Quinoline yellow	Cold fruits, ice creams, cake, chocolate, bread, cheese sauces and beverage products
		Sunset yellow	Jam, dairy products, cocoa products, starch desserts, compound condiments, beverage products
		Carmoisine	Beverage, wine, candy, green plum, bayberry, sandwich, ice cream
		Ponceau 4R	Beverage, wine, soda, candy, pastry, soya drink, ice cream, yogurt
		Allura red	Candy coating, fried chicken, meat enema, western ham, jelly, biscuit sandwich
Preservatives	Natural	Nisin	Meat, dairy products, vegetable protein products, canned goods, coffee beverage, tea, soy sauce
		Natamycin	Yogurt, cheese, raw ham, dried sausage, cakes
		Lysozyme	Cheese, drink, baby food, meat and fish products
	Artificial	Sorbic acid and potassium sorbate	Dairy products, soybean products, processed vegetables, cooked meat products, aquatic products
		Benzoic acid and sodium benzoate	Condiments, pickled products, beverage products, fruit wine
		p-hydroxybenzoates	Jam and sauce products, carbonated drinks
		Sodium nitrite	Soybean products, meat products, aquatic products, pastry and puffed foods
		Propionic acid and sodium propionate	Soybean products, wet flour products, bread, pastry, vinegar and soy sauce
Antioxidants	Natural	Ascorbic acid and its salts	Peeled fresh fruits and vegetables, wheat flour, fruits and vegetables products
	Artificial	Tocopherol and its geometric isomers	Meat, fish, nuts, vegetables, fruits, beverages and canned food
		Propyl gallate	Nuts and canned seeds, gum-based candy, grilled meat, fried noodles
		Tert-butyl hydroquinone	Moon cakes, instant rice noodles products, biscuits, baked food fillings
		Butyl hydroxyanisole	Fat, oil and emulsified fat products, coarse grain, instant rice noodles products
		Butylated hydroxytoluene	Fried noodles, gum-based candy, air-dried aquatic products

(continued)

Category		Common substance	Mainly used
Sweeteners	Natural	Sorbitol	Dairy products, jams, wet flour products, baked products, beverages and soybean products
		Mannitol	Candy, chewing gum
		Thaumati	Candied fruits, candy, biscuits, canned meat
		Stevioside	Flavor fermented milk, candy, condiments, canned fruits, flavored sirup and tea products
		Maltitol, maltitol sirup	Processing fruits, frozen surimi products, soybean products
		Lactitol	Dairy products, spices
		Xylitol	Dairy products, tea products, alcoholic drinks, seasonings, starch products, processed fruits and vegetables
	Artificial	Aspartame	Dairy products, frozen fruits and vegetables, cereals and starch desserts
		Cyclamate	Canned fruits, jams, mixed wine, instant noodle food and condiments
		Saccharin and its salts	Frozen drinks, dehydrated mango, dried figs, cold fruits, cooked beans and dried fruits
		Sucralose	Prepared dairy products, jams, sufu, coarse cereals products, Baked products
		Neotame	Dairy products, frozen fruits and vegetables, cereals and starch desserts
Thickeners		Agar	Ice cream, low-fat spreads, dairy products, salad dressings, mayonnaise
		Pectin	Jam, jelly, cheese, candy, sauce, yoghurt
		Gellan gum	Pudding, jelly, sugar, drinks, dairy products, jam products, bread

Colorants

Colorants are a kind of additives that added to food to remedy colors lost during preparation, enhance perception of flavor or to make food look more attractive. Colorants present in both natural and synthetic forms, where natural food colors refer to naturally available dyes obtained from vegetables, animals or minerals. For instance, curcumin is mainly used for coloring food like beverages, sauces and confectionery, which can be added only lower than 0.01 g/kg. On the other hand, synthetic food colors are popularly also known as artificial food colors, which are manufactured chemically. Wherein, quinoline yellow used for coloring cold fruits, ice creams, sweets, and beverage products should be controlled under 0.1 g/kg in food processing.

Preservatives

Preservatives are used to prevent or inhibit spoilage of food due to contamination of fungi, bacteria and other microorganisms and keep food safe for longer time. Similar to food colors, preservatives can also be divided into two main groups of artificial and natural ones. Commonly used artificial preservatives include sorbic acid, which is used in dairy products, soybean products, processed vegetables, cooked meat products and aquatic products. The natural preservatives involving nisin, which is added to meat, dairy products, vegetable protein products, canned goods, coffee beverage, tea and soy sauce.

Antioxidants

Antioxidants mainly prevent or inhibit the oxidation process to keep the food from spoilage. Natural antioxidant additives are commonly added to meat, fish, nuts, vegetables, fruits, beverages and canned food. For example, the additive amount of ascorbic acid as well as its salt-like ascorbate and its geometric isomers should be controlled from 0.2 to 5.0 g/kg. The artificial antioxidants are often added to oils, cheese, and chips to suppress the formation of hydroperoxides. For instance, the phenol derivatives of propyl gallate should be controlled in dosage from 0.1 to 0.4 g/kg.

Sweeteners

Sweetener, a sugar substitute, is a food additive that provides a sweet taste like that of sugar while containing significantly less food energy, making it a zero-calorie or low-calorie sweetener. Natural sweeteners like sorbitol, xylitol, and mannitol are derived from sugars, which are widely used in dairy products, tea products, alcoholic drinks, seasonings, candy, starch products, processed fruits and vegetables. Artificial sweetener such as aspartame, cyclamate, saccharin sodium and sucralose are mainly used in fruits, jams, beverages, desserts and dairy products.

Thickeners

Thickeners or thickening agents are substances added to food preparations for increasing their viscosity without changing other properties like taste, for example, pectin has the maximum dosage of lower than 3.0 g/kg. Although food additives bring so many benefits, the improper use of food additives will do harm for human health.

However, these additives have brought other unwanted outcomes and are an issue of concern. Despite all the benefits and advantages of food additives and preservatives, there is still a potential danger of chemical adulteration of foods. Additives or preservatives in foods may themselves trigger other hormonal or chemical processes in the body that can generate negative physiological responses. The metabolites produced by additives may also cause side effects, because not all food additives enter the markets after being thoroughly studied to prove their safety. Although most food additives are considered safe, some are known to be carcinogenic or toxic. For these reasons, many food additives and preservatives are controlled and regulated by national and international health authorities. All food manufacturers must comply with the standards set by the relevant authorities without violating the maximum thresholds stated to ensure the safety of the final product to the consumers. In most cases, food processing industries must seek standard certification before using any new additive or preservative or before using any originally certified additive or preservative in a different way.

New Words

defoaming agent		消泡剂
stabilizer ['steɪbəlaɪzə(r)]	*n.*	稳定剂
coagulant [kəʊ'ægjʊlənt]	*n.*	凝固剂；促凝剂
sweetener ['swiːtnə(r)]	*n.*	甜味剂
thickener ['θɪkənə(r)]	*n.*	增稠剂
dye [daɪ]	*n.*	染料
curcumin ['kɜːkjʊmɪn]	*n.*	姜黄素
quinoline yellow		喹啉黄
preservative [prɪ'zɜːvətɪv]	*n.*	防腐剂；预防法；预防药
	adj.	保存的；有保存力的；防腐的
sorbic acid		山梨酸
nisin ['naɪsɪn]	*n.*	乳酸链球菌肽；尼生素
geometric [dʒiːə'metrɪk]	*adj.*	几何学的
phenol ['fiːnɒl]	*n.*	苯酚；酚类；酚类化合物
derivative [dɪ'rɪvətɪv]	*n.*	派生物；衍生字；导数
	adj.	派生的；（贬）非独创的；庸乏的
propyl ['prəʊpəl]	*n.*	丙烷基
gallate ['gæleɪt]	*n.*	没食子酸盐
carotene ['kærətiːn]	*n.*	胡萝卜素
anthocyanin [ˌænθə'saɪənɪn]	*n.*	花青素
canthaxanthin [kæn'θæksænθɪn]	*n.*	角黄素；斑蝥黄
tartrazine ['tɑːtrəziːn]	*n.*	柠檬黄
sunset yellow		日落黄
carmoisine [kɑːm'wɑːsiːn]	*n.*	胭脂红
natamycin [neɪtə'maɪsɪn]	*n.*	纳他霉素
lysozyme ['laɪsəzaɪm]	*n.*	溶菌酶

sorbic acid and potassium sorbate *n.* 山梨酸和山梨酸钾

benzoic acid and sodium benzoate *n.* 苯甲酸和苯甲酸钠

p-hydroxybenzoates *n.* 对羟基苯甲酸酯；尼泊金酯类

sodium ['səʊdiəm] nitrite 亚硝酸钠

propionic acid and sodium propionate *n.* 丙酸和丙酸钠

tocopherol [tɒ'kɒfəˌrɒl] *n.* 维生素 E；生育酚

tert-butyl hydroquinone *n.* 叔丁基对苯二酚

butyl hydroxyanisole *n.* 丁基羟基苯甲醚

butylated hydroxytoluene *n.* 丁基羟基甲苯

sorbitol ['sɔːbɪtəl] *n.* 山梨醇

mannitol [mæ̀nɒl] *n.* 甘露醇

thaumati [θɔː 'mɑːtɪ] *n.* 非洲竹芋甜素

stevioside ['stiːvɪəsaɪd] *n.* 甜菊糖，甜菊苷，蛇菊苷（一种非营养性的天然甜味剂，比蔗糖甜 300 倍）

maltitol ['mɔːltɪtəl] *n.* 麦芽糖醇

sirup ['sɪrəp] =syrup *n.* 糖浆

lactitol ['læktɪtɒl] *n.* 乳糖醇

xylitol ['zaɪlɪtɒl] *n.* 木糖醇

aspartame [ə'spɑːteɪm] *n.* 阿斯巴甜

cyclamate ['saɪkləmeɪt] *n.* 甜蜜素

saccharin ['sækərɪn] *n.* 糖精

sucralose ['suːkrələʊs] *n.* 三氯蔗糖

neotame [niə'teɪm] *n.* 纽甜

agar ['eɪgɑː(r)] *n.* 琼脂

pectin ['pektɪn] *n.* 果胶

gellan gum 葛兰胶

confectionery [kən'fekʃənəri] *n.* 甜食（糖果、巧克力等）

jelly ['dʒeli] *n.* 果冻；胶状物
vi. 结冻；做果冻
vt. 使结冻

carbonated drinks 碳酸饮料

soya drink 大豆饮料

biscuit sandwich 饼干三明治

dried sausage 风干肠

aquatic [ə'kwætɪk] *adj.* 水生的；水中的；水上的
n. 水生动物；水草

condiment ['kɒndɪmənt]	*n.*	调味品
pickled ['pɪkld]	*adj.*	腌渍的
surimi [sjuː 'rɪmaɪ]	*n.*	鱼糜
spices [s'paɪsɪz]	*n.*	香味料；调味料；趣味，情趣
seasoning ['siːzənɪŋ]	*n.*	调料品
dehydrated [ˌdiːhaɪ'dreɪtɪd]	*v.*	使（食物）脱水
sufu ['suːfʊ]	*n.*	腐乳
sauce [sɔːs]	*n.*	调味汁；酱
	vt.	给…调味；使…增加趣味；对…无礼
pudding ['pʊdɪŋ]	*n.*	布丁
improper [ɪm'prɒpə(r)]	*adj.*	不合适的；错误的；不道德的
trigger ['trɪgə(r)]	*v.*	触发；引起
	n.	扳机；起因
hormonal [hɔː 'məʊnl]	*adj.*	激素的；荷尔蒙的
carcinogenic [ˌkɑːsɪnə'dʒenɪk]	*adj.*	致癌的；致癌物的

Notes

1) While chemically synthesized additives are based on chemical raw materials, from which organic or inorganic matter can be extracted and purified.
参考译文：而化学合成添加剂是以化学原料为基础，从中提取和纯化有机或无机物质作为添加剂。
句中“While…”表示与上一句相对，理解为“而…”。“which”指代“chemical raw materials”。
2) Sweetener, a sugar substitute, is a food additive that provides a sweet taste like that of sugar while containing significantly less food energy, making it a zero-calorie or low-calorie sweetener.
参考译文：甜味剂是一种替代糖的食品添加剂，它能具有像糖一样的甜味。但其所含的能量显著低于糖，这使其成为零卡路里或低卡路里呈甜味物质。
句中“a sugar substitute”是对“sweetener”的解释，“that”引导的定语修饰“food additive”，“making it…”状语从句，理解为“使其…”。

Lesson 2 Food Enzymes and Its Application

The presence of enzymes in nature have been well known for over a century. Enzymes, also known as biocatalysts, are a biological substance that initiates or accelerates the rate of a biochemical reaction in a living organism, without itself being consumed in the reaction. Even though enzymes are produced inside the living cells, they can work actively *in vitro*, making them useful in industrial processes. The literature suggested that >5,000 biochemical reaction types are catalyzed by the enzymes. Similar to other chemical catalysts, enzymes are also highly effective in increasing the rate of biochemical reactions that otherwise proceed very slowly, or in some cases, not at all. A common example is the breakdown of foods, which includes mainly proteins, carbohydrates and fats, into their basic constituents. It is normally accomplished within 3-6 h depending on the type and amount of food. However, in the absence of enzymes, this breakdown of foodstuffs would take >30 years. In comparison to chemical catalysts, enzymes are more specific in action and possess high catalytic properties. Also, enzymes can be immobilized on inert support material without loss of activity that facilitates their reuse and recycling.

Most enzymes, but not all, require a small molecule to perform their activity as a catalyst. These molecules are known as cofactors or coenzymes. Cofactors are non-proteinaceous chemical compounds that are bound to an inactive protein part of enzyme (apoenzyme) in order to increase the biological activity of the enzyme required for its function. The active complex of apoenzyme (protein part) along with cofactor (coenzyme or prosthetic group) is referred to as holoenzyme. Cofactor is also considered a "helper molecule" because it assists in biochemical transformations. There are two types of cofactors: coenzymes and prosthetic groups. Coenzymes are a specific type of cofactor and are organic molecules that bind to enzymes and help in their functions. Prosthetic groups (organic molecules or metal ions) are also cofactors that often bind tightly to proteins or enzymes by a covalent bond (Figure 4.1).

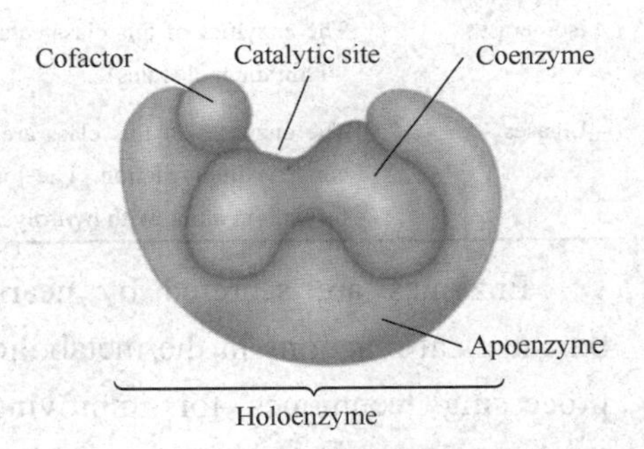

Figure 4.1 Composition of enzyme

To date, >6,000 different types of enzymes are known. The names of commonly used enzymes are based on the type of reaction they catalyze followed by the suffix-ase. For example, the hydrolysis of proteins is catalyzed by proteases. There are also some trivial names for the initially studied enzymes such as trypsin, pepsin, rennin, etc. However, trivial names give no indication of source, function or reaction catalyzed by the enzyme. Thus a variety of different names have been used for the same enzyme and different enzymes were known by the same name. Also, due to lack of consistency in the nomenclature and rapid growth in enzyme discovery, there was a need for a systematic way to name and classify enzymes. The International Union of Biochemists (IUB) developed an unambiguous system of enzyme nomenclature in which each enzyme had a unique name and four-digit code number, prefixed by EC and separated by points that identify the substrate acted upon and the type of reaction catalyzed. These four-digit code numbers, prefixed by EC, provide the following information: (a) the first number of EC code indicates six main classes of the enzyme, (b) the second number of EC code denotes subclass, (c) the third number indicates sub-subclass, and (d) the fourth number of EC code is the serial number of the enzyme in its sub-subclass. This is now in widespread use and the approved list of classified enzymes can be found at http://www.enzyme-database.org. Enzymes are classified into six main classes based on the type of chemical reaction catalyzed: Oxidoreductases (EC 1.×.×.×), Transferases (EC 2.×.×.×), Hydrolases (EC 3.×.×.×), Lyases (EC 4.×.×.×), Isomerases (EC 5.×.×.×) and Ligases (EC 6.×.×.×) (Table 4.2).

Table 4.2 Six classes enzymes and the type of chemical reaction catalyze

Class	Type of chemical reaction catalyzed	Example
Oxidoreductases	The enzymes of this class are involved in redox reactions in which hydrogen or oxygen atoms or electrons are transferred between molecules	Glucose oxidase (EC 1.1.3.4)
Transferases	The enzymes of this class catalyze transfer of specific functional groups (e.g., alkyl-, glycosyl-) between two molecules, but excluding oxidoreductases and hydrolases	Aspartate aminotransferase (EC 2.6.1.1)
Hydrolases	The enzymes of this class are involved in hydrolytic reactions and their reversal (use water to cleave chemical bonds)	Alkaline phosphatase (EC 3.1. 3.1)
Lyases	The enzymes of this class are involved in elimination reactions. This is nonhydrolytic removal of a group of atoms from the substrate	Histidine carboxy-lyase (EC 4.1.1.22)
Isomerases	The enzymes of this class catalyze molecular isomerization (transfer of groups within the molecules)	Xylose isomerase (EC 5.3.1.5)
Ligases	The enzymes of this class are also known as synthetases and are involved in condensation reaction. The joining of two molecules involves covalent bond formation, along with hydrolysis of a nucleoside triphosphate	Glutathione synthase (EC 6.3.2.3)

Enzymes are secreted by nearly all living cells for catalysis of their own specific biochemical reactions in the metabolic process. Enzymes are playing an important role in food processing techniques for improving nutritive value and flavor of processed food. The food-processing industry—the making of cheese, leavened bread, wine and beer, yogurt, and syrup—is successfully using enzymes at the commercial level.

Glucose Oxidase

The glucose oxidase enzyme is commercially produced from *Aspergillus niger* and *Penicillium glaucum* through a solid-state fermentation method. Fungal strains *A. niger* are able to produce notable amounts of glucose oxidase. Glucose oxidase enzymes are used to remove small amounts of oxygen from food products or glucose from diabetic drinks. Glucose oxidase is playing an important role in color development, flavor, texture, and increasing the shelf life of food products.

Laccase

Laccase is responsible for discoloration, haze, wine stabilization, baking, and flavoring in food processing. Laccase improves the baking process through an oxidizing effect and provides an additional development in the strength of dough and baked products, including enhancing crumb structure and increasing softness and volume. Another diverse application of laccase is in environmental sectors, which degrade various ranges of xenobiotic compounds.

Transglutaminase

Transglutaminase is responsible for acyl transfer, deamidation, and the inter- and intra-molecular crosslink between amino acid residues of glutamine and lysine. The commercial application of transglutaminase enzymes in the food-processing industry is improving the protein-emulsifying capacity, gelatation, viscosity, and production of various types of protein ingredients to enhance the quality of food products. Transglutaminase is enhancing the water-holding capacity, softness, foam formation, and stability of food products.

Lactase

Lactase enzymes catalyze the breakdown of the milk sugar lactose into simple sugar monomer units like glucose and galactose. Lactases are obtained from plants, animal, bacteria, fungus, yeasts, and molds. Commercial production of lactase enzymes are developed from *A. niger*, *A. oryzae*, and *Kluyveromyces lactis*. Fungal origin lactases have optimum activity at acidic pH ranges, and yeast and bacterial-originated lactases have optimum pH ranges near to neutral. The lactase enzyme is predominantly rich in infancy and is called a brush border enzyme. Some people do not produce enough of the lactase enzyme so they do not properly digest milk. This is called lactose intolerant, and people who are lactose intolerant need to supplement the lactase enzyme to aid in the digestion of milk sugar. Another useful application of the lactase enzyme is it increases the sweetness of lactase-treated milk, and assists in the manufacturing of ice cream and yogurt preparation.

Catalase

Catalase enzymes break down hydrogen peroxide (H_2O_2) to water and oxygen molecules, which protects cells from oxidative damage by reactive oxygen species. Commercial catalases are produced from *A. niger* through a solid-state fermentation process. The major applications of catalase in the food-processing industry include working with other enzymes like glucose oxidase, which is useful in food preservation and egg processing, and sulphydryl oxidase, which under aseptic conditions, can eliminate the effect of volatile sulphydryl groups, that is, they generate from thermal induction and are responsible for the cooked/off-flavor in ultra-pasteurized milk.

Lipase

Lipases catalyze the hydrolysis of ester bonds in lipid substrates and play a vital role in digestion and the transport and processing of dietary lipids substrate. Lipases catalyze the biochemical reaction like esterification, interesterification, and transesterification in nonaqueous media which frequently hydrolyze triglycerides into diglycerides, monoglycerides, fatty acids, and glycerol. Microorganism like *Pseudomonas aeruginosa*, *Serratia marcescens*, *Staphylocococcus aureus*, and *Bacillus subtilis* are the best sources of lipase enzymes. Lipases are widely used in pharmacological, chemical, and food industries. The commercial applications of lipases in the food industry are the hydrolysis of milk fats, pronounced cheese flavor, low bitterness, and prevention of rancidity. Lipases may combine with many other enzymes like protease or peptidases to create good cheese flavor with low levels of bitterness.

Protease

Proteolytic enzymes are also termed as peptidases, proteases, and proteinases, which are able to hydrolyze peptide bonds in protein molecules. Proteases are generally classified as endopeptidases and exopeptidases. Exopeptidases cut the peptide bond proximal to the amino or carboxy termini of the protein substrate, and endopeptidases cut peptide bonds distant from the termini of the protein substrate. Proteases are obtained from diverse groups of organisms such as plants, animals, and microorganisms, but commercially viable proteases are obtained from microorganisms, especially bacterial and fungal species. Broad working range of temperature (10-80℃) and pH (4-12) of protease enzymes increases their application in the food-processing industry, the major role in cheese and dairy product manufacturing. Aminopeptidases are significantly improving the flavor in fermented milk products. Other basic applications of proteases in the food-processing industry are to increase the nutritive value of bread, baked goods, and crackers.

α-Amylase

Amylase enzymes hydrolyze complex starch molecules into simple monomer units of glucose. Sources of α-amylase are plants, animals, and microorganisms, but commercially viable amylases are produced from microorganisms, especially bacterial and fungal species. Starch-converting properties of α-amylases are playing an important role in the food, beverage, and sugar industries. α-Amylase is improving the quality of breads that have reduced size and poor crust color, and compensates for the nutritional deficiencies of the grain. α-Amylase also degrades the starch in wheat flour into small dextrins, thus allowing yeast to work continuously during dough fermentation, proofing, and the early stages of the baking process. α-Amylases are also employed in many other aspects of the food industry like clarification of beer, fruit juices, and pretreatment of animal feed to improve the digestibility of fiber.

Xylose (Glucose) Isomerase

Xylose isomerase (d-xylose ketol-isomerase) catalyzes the isomerisation reaction of D-xylose into xylulose. The greatest application for glucose isomerase is in the food-processing industry; it mainly catalyzes two significant reactions such as reversible isomerization of d-glucose to d-fructose, and d-xylose to d-xylulose.

New Words

biocatalyst [bi'ːəʊkətəlɪst]	*n.*	生物催化剂
initiate [ɪ'nɪʃieɪt]	*vt.*	开始；创始；启蒙；介绍加入
	n.	创始人
	adj.	新加入的；启蒙的
in vitro		在生物体外
holoenzyme [hɒləʊ'enzaɪm]	*n.*	全酶
proteinaceous [prəʊtiː 'neɪʃəs]	*adj.*	蛋白质的
apoenzyme [ə'pəʊnziːm]	*n.*	脱辅基酶；主酶；酶朊；脱辅基酶蛋白
suffix ['sʌfɪks]	*n.*	后缀
	vt.	添后缀
trivial ['trɪviəl]	*adj.*	琐碎的；无价值的
trypsin ['trɪpsɪn]	*n.*	胰蛋白酶
pepsin ['pepsɪn]	*n.*	胃蛋白酶
rennin ['renɪn]	*n.*	凝乳酶（用来制干酪和凝乳食品）
consistency [kən'sɪstənsi]	*n.*	连贯性；一致性；强度；硬度；浓稠度
nomenclature [nə'menklətʃə(r)]	*n.*	系统命名法
unambiguous [ˌʌnæm'bɪgjuəs]	*adj.*	不含糊的
prefixed ['priːfɪkst]	*n.*	前缀；（人名前的）称谓
	vt.	加…作为前缀；置于前面
oxidoreductase ['ɒksɪdəʊrɪ'dʌkteɪs]	*n.*	氧化还原酶
transferase ['trænsfəreɪs]	*n.*	转移酶
hydrolase ['haɪdrəleɪs]	*n.*	水解酶
lyase ['laɪəs]	*n.*	裂合酶；裂解酶
isomerase [aɪ'sɒməreɪs]	*n.*	异构酶

ligase [lɪ'geɪs]	*n.*	连接酶
aspartate [ə'spɑrteɪt]	*n.*	天（门）冬氨酸盐
aminotransferase [æmaɪnətrænsf'reɪz]	*n.*	转氨酶
histidine carboxylyase		组氨酸羧化酶
triphosphate [traɪ'fɒsfeɪt]	*n.*	三磷酸盐
glutathione [glu:tə'θaɪəʊn]	*n.*	谷胱甘肽
non hydrolytic		非水解的
diabetic [ˌdaɪə'betɪk]	*adj.*	糖尿病的
	n.	糖尿病患者
discoloration [disˌkʌlə'reiʃn]	*n.*	变色；褪色；（皮肤上的）疹斑
laccase ['lækeɪs]	*n.*	漆酶；虫漆酶
haze [heɪz]	*n.*	烟雾
	vi.	变糊涂；变糊涂
	vt.	戏弄
xenobiotic [ˌzenəʊbaɪ'ɒtɪk]	*n.*	异型生物质
	adj.	异型生物质的
transglutaminase [trænzglu: 'tæmɪneɪs]	*n.*	转谷氨酰胺酶
deamidation [di:æmɪ'deɪʃən]	*n.*	脱酰胺（作用）
glutamine ['glu:təmi:n]	*n.*	谷氨酰胺
gelatinization	*n.*	凝胶化
lactase ['lækteɪs]	*n.*	乳糖酶
catalase ['kætəleɪs]	*n.*	过氧化氢酶
sulphydryl [sʌl'faɪdrɪl]	*n.*	巯基；氢硫基
ultra-pasteurized		超巴氏杀菌的
esterification [eˌsterɪfɪ'keɪʃən]	*n.*	酯化作用
interesterification [ɪntərestərɪfɪ'keɪʃən]	*n.*	酯交换；相互酯化
transesterification [trænsəsterəfɪ'keɪʃən]	*n.*	酯基转移
nonaqueous ['nɒn'eɪkwɪəs]	*adj.*	非水的
diglycerides [daɪ'glɪsəˌraɪdz]	*n.*	甘油二酯
monoglycerides [mɒ'nəgli:səraɪdz]	*n.*	甘油一酸酯
pharmacological [ˌfɑ:məkə'lɒdʒɪkl]	*adj.*	药理学的
protease ['prəʊtieɪz]	*n.*	蛋白酶
peptidase ['peptɪdeɪs]	*n.*	肽酶
endopeptidase [endəʊ'peptɪdeɪs]	*n.*	肽链内切酶
exopeptidase [eksəʊ'peptɪdeɪs]	*n.*	肽链端解酶；外肽酶

termini ['tɜːmɪnaɪ]	*n.*	目的地；界标；终点（terminus 的名词复数）
aminopeptidases [æmɪnoʊ'peptaɪdeɪz]	*n.*	氨基肽酶
amylase ['æmɪleɪz]	*n.*	淀粉酶

Notes

1) Enzymes, also known as biocatalysts, are a biological substance that initiates or accelerates the rate of a biochemical reaction in a living organism, without itself being consumed in the reaction.

   参考译文：酶（生物催化剂）是一种能够启动或加速生物体内的生化反应速率的生物物质，而酶自身不会在反应中消耗。

   句中"that"引导的定语从句修饰"biological substance"，"without…"是介词短语表示状态。

2) Similar to other chemical catalysts, enzymes are also highly effective in increasing the rate of biochemical reactions that otherwise proceed very slowly, or in some cases, not at all.

   参考译文：与其他化学催化剂类似，酶在提高生化反应速率方面也非常有效，否则生化反应进行得非常缓慢，或者在某些情况下根本无法进行。

   句中"Similar to"表示状态，"that"引导的定语从句修饰"reactions"。

Reading Material 1 Regulations of Food Additives

Food additives regulation in countries with existing procedures agrees with the general principle: (a) that food additive safety can be reasonably assured by critically designed animal studies, (b) that the determination of safe level should be based on maximum dietary level producing no adverse effect in test animals, (c) that the intake of the additive will be below that which could produce harmful effects in animals, (d) that adjustment should be made to account for the safety of vulnerable populations, and (e) that the determination of safety must be based on the judgment of scientists qualified to render such determination. There is also universal acceptance that, for a major of new food additive, adequate animal studies are necessary to address potential mutagenicity (*n.* 诱变性), subchronic (*adj.* 亚慢性) and chronic toxicity, reproductive and developmental toxicity, and carcinogenicity (*n.* 致癌力) at a minimum.

Harmonization (*n.* 协调) of food additive regulations, however, is an elusive (*adj.* 难以达到的) goal because of major differences in global food use patterns, in the definitions of various additives and in current regulations. For example, the first major difference is that the only country with a GRAS list is the US. This means that compounds considered GRAS in the US may still need formal approvals in other countries. In a way, Japan has an informal GRAS approach in that natural products, either in plants or through fermentation, are considered inherently safe. Thus, a natural compound that has undergone little testing in Japan but has been used safely in the Japanese population for years could require investigation if exported to the US or to the European Union (EU). China considers nutrition enhancers, gum-based substances in chewing gum and flavoring agents as direct food additives whereas other countries do not. The regulations of food additives vary among government agencies and organizations and the regulations of Codex Alimentarius (国际食品法典委员会), China, the EU, the US, Japan and Indian are listed in below.

Codex Alimentarius

General Standard for Food Additives Codex STAN 192-1995 was adopted in 1995 and subsequently revised in 1997, 1999, 2001, 2003, 2004, 2005, 2006, 2007, 2008, 2009, 2010, 2011, 2012, 2013, 2014, 2015, 2016, 2017, 2018 and 2019. The Joint Expert Committee on Food Additives (JECFA) of FAO/WHO is a scientific body that evaluates the safety of food additives, processing aids, flavorings, residues of veterinary drugs in animal products, contaminants and natural toxins, exposure assessments, methods and development of general principles. JECFA reviews the safety of an additive or group of additives and based on available scientific safety data, develops a specification that includes description of the additive, its functional uses in food, identity, purity and methods used to characterize the additives. An additive must have JECFA specifications confirming its safety and INS number for its inclusion in General Standard for Additives (GSFA) list. The listed additives are then recognized within Codex Committee of Food Additives (CCFA).

The Codex standard on food additives defines food additives and other related terms as follows:

(1) Food additive means any substance not normally consumed as a food by itself and not normally used as a typical ingredient of the food, whether or not it has nutritive value, the intentional addition of which to food for a technological (including organoleptic) purpose in the manufacture, processing, preparation, treatment, packing, packaging, transport or holding of such food results, or may be reasonably expected to result (directly or indirectly), in it or its by-products becoming a component of or otherwise affecting the characteristics of such foods. The term does not include contaminants or substances added to food for maintaining or improving nutritional qualities.

(2) Acceptable daily intake (ADI) is an estimate by JECFA of the amount of a food additive, expressed on a body weight basis that can be ingested daily over a lifetime without appreciable health risk. The ADI is expressed in milligrams of the chemical, as it appears in the food, per kilogram of body weight per day.

(3) ADI "Not Specified" (NS) is a term applicable to a food substance of very low toxicity

for which, on the basis of the available data (chemical, biochemical, toxicological, and other), the total dietary intake of the substance, arising from its use at the levels necessary to achieve the desired effect and from its acceptable background levels in food, does not, in the opinion of JECFA, represent a hazard to health. For the above reason, and for reasons stated in individual JECFA evaluations, establishment of an ADI expressed in numerical form is not deemed necessary by JECFA. An additive meeting the above criterion must be used within the bounds of Good Manufacturing Practice (GMP).

(4) Maximum Use Level of an additive is the highest concentration of the additive determined to be functionally effective in a food or food category and agreed to be safe by the Codex Alimentarius Commission. It is generally expressed as mg additive $\cdot$ kg^{-1} of food. The Maximum Use Level will not usually correspond to the optimum, recommended, or typical level of use. Under GMP, the optimum, recommended, or typical use level will differ for each application of an additive and is dependent on the intended technical effect and the specific food in which the additive would be used, taking into account the type of raw material, food processing and post-manufacture storage, transport and handling by distributors, retailers, and consumers.

(5) Processing aid means any substance or material, not including apparatus or utensils, and not consumed as a food ingredient by itself, intentionally used in the processing of raw materials, foods or its ingredients to fulfill a certain technological purpose during treatment or processing and which may result in the non-intentional but unavoidable presence of residues or derivatives in the final product.

The scope of this standard is explained in four different parts which include:

(1) Only the food additives, listed in this standard with assigned ADI as well as technologically justified, are recognized as suitable for their use in foods.

(2) Codex commodity committees have the responsibility and should technologically justify the use of additives in various foods.

(3) Food categories in which the use additives are not acceptable, are also defined by this standard.

(4) Maximum use levels for food additives should be established keeping in view the ADI.

China

The functional range of food additives is the largest in Chinese food regulation, which defines food additives as both chemosynthetic (*adj.* 化学合成的) and natural substances added to food for quality, color, fragrance and taste improvement or for food preservation and processing technology. Therefore, coloring agents, sweeteners, flavoring agents, preservatives, antioxidants, processing acids, nutritive additives, gum-based substances all belong to food additives. There are 22 allowed food additive categories and more than 2,000 varieties food additives in China at present. As an essential material in food industry, the quality and safety of food additives are also of concern. With the implementation of GB 2760—2014 and GB 29924—2013, the requirements for food additives usage and their Chinese label are more comprehensive.

In China, food additives are mainly regulated by the following standards and laws:

(1) Administrative Measure on New Food Additives.

(2) Provision on New Food Additives Application and Acceptance.

(3) GB 2760—2014 The Usage of Food Additives.

(4) GB 28050—2012 The Usage of Nutrient Enhancer.

(5) GB 29987—2014 Gum Base and its Ingredients.

(6) GB 29924—2013 General Rules for the Labelling of Food Additives.

(7) GB 29938—2013 General Rules for Food Spices.

(8) GB 26687—2011 General Rules for Compound Food Additives.

(9) GB 30616—2014 Food Flavors.

(10) GB 14880—2012 Standard for Use of Food Nutrition Fortifiers.

(11) GB 7718—2011 General Rules for Labeling of Prepackaged Food.

Food additives are allowed to be used in China only if they are: covered by national food safety standards GB 2760—2014 and GB 14880—2012; within the list of allowable food additives announced by the Ministry of Health (MOH); and within the scope of allowed applications and dosage levels. It is very important to determine whether the additive is new or not in China. If the additive is new, registration of the additive with MOH is required.

Food Additive shall be clearly presented in the prominent place of a label in China. The names of food additives must be consistent with GB 2760—2014 or GB 14880—2012 or the notice of MOH. Each additive shall be declared in a descending order of the content of each ingredient. The scope of use and the allowable dosage of a food additive as well as its application method shall also be given. In case of compound additives, the quantity of each food additive shall be indicated in a descending order. However, the content of each ingredient for a compound food additive does not need to be given in case of non-retail sales of food additive. When food additives are used in prepackaged food for direct delivery to consumers, they shall be indicated on the label in descending order of their weights added in the process of manufacture or preparation of the food. The names of those food additives shall be declared in general names in accordance with GB 2760—2014. The content of each ingredient does not need to be declared. This labeling requirement is specified in GB 7718—2011.

The EU

According to the EU legislations [Regulation (EC) No 1333/2008 of the European Parliament and of the Council of December 16, 2008 on Food Additives], food additive has been defined as "any substance not normally consumed as a food in itself and not normally used as a characteristic ingredient of food, whether or not it has nutritive value", the intentional addition of which to food for a technological purpose in the manufacture, processing, preparation, treatment, packaging, transport or storage of such food results, or may be reasonably expected to result, in it or its by-products becoming directly or indirectly a component of such foods. EU legislation has defined 26 classes of "technological purposes" such as color (used to add or restore color in a food), preservatives (added to prolong the shelf-life of foods by protecting them against micro-organisms), antioxidants (substances, which prolong the shelf-life of foods by protecting them against oxidation), flour treatment agents (added to flour or to dough to improve its baking quality), etc. The safety of all food additives that were authorized had been assessed by the Scientific Committee on Food (SCF) and/or the European Food Safety Authority (EFSA). Only additives for which the proposed uses were considered safe are on the EU list.

The regulation 1333/2008 specifically excludes:

(1) Processing aids.

(2) Substances used for the protection of plants and plant products in accordance with community rules relating to plant health (e.g., pesticides, herbicides, insecticides).

(3) Substances added to foods as nutrients (e.g., minerals or vitamins).

(4) Substances used for the treatment of water for human consumption falling within the scope of Council Directive 98/83/EC on drinking water quality.

(5) Flavorings as they are regulated under Regulation (EC) 1334/2008 on flavorings and certain food ingredients with flavoring properties.

(6) Food enzymes as they are controlled under Regulation (EC)1332/2008 on food enzymes.

Extraction solvents are also not considered as food additives in the EU and are subject to specific legislation on both their use and residual levels, under Directive 2009/32/EC.

The US

The US definition of a food additive is very different from the EU and Codex and is quite unique. The Federal Food, Drug and Cosmetic Act (FFDCA) entrusts authority on FDA to regulate the safety of food, drugs, medical devices and cosmetics. The current legislation on food additives of the US is based on the Food Additives Amendment, which was enacted to the FFDCA in 1958. FDA is the primary food regulatory authority of the US which is a part of HHS. FDA have responsibility of nearly all the foods which are produced, marketed, sold and consumed in the US including food ingredients and food additives. It defines the terms "food additive" and "unsafe food additive" and established a premarket approval process for food additives. Food additives includes all substances not exempted by section 201(s) of the act, the intended use of which results or may reasonably be expected to result, directly or indirectly, either in their becoming a component of food or otherwise affecting the characteristics of food (including any substance intended for use in producing, manufacturing, packing, processing, preparing, treating, packaging, transporting or holding food; and including any source of radiation intended for any such use). If such substance is not generally recognized, among experts qualified by scientific training and experience to evaluate its safety, as having been adequately shown through scientific procedures (or, in the case of a substance used in food prior to January 1, 1958, through either scientific procedures or experience based on common use in food) to be safe under the conditions of its intended use. Further, a substance that does not become a component of food, but that is used, for example, in preparing an ingredient of the food to give a different flavor, texture, or other characteristic in the food, may be a food additive.

The above definition excludes:

(1) Color additives (controlled by separate provisions in the FFDCA).

(2) Pesticide chemical.

(3) Pesticide chemical residue in or on a raw agricultural commodity or processed food.

(4) Any substance used in accordance with a sanction or approval granted prior to September 6, 1958.

Therefore, in the US, three following broad categories of substances intended for use directly or indirectly, in the manufacture of foodstuffs for human consumption without adulterating the food are:

(1) Prior-sanctioned (事先批准) food ingredients [substances that received official approval for their use in food by the FDA or the USDA prior to the Food Additives Amendment in 1958] and described in the 21 CFR 181 Prior sanctioned food substances means approval of a substance and its use in or on food for which the approval has been granted by FDA or USDA prior to September 6, 1958. A prior sanction shall exist only for a specific use of a substance in food for which there was explicit approval by the FDA and USDA. For example, USDA's sanction to use nitrate and nitrite (*n.* 硝酸盐和亚硝酸盐) in curing of meat and poultry (*n.* 家禽).

(2) Substances that are GRAS and described in 21 CFR 182-184. Regardless of whether the use of a substance is a food additive or in GRAS, there must be evidence that the substance is safe under the conditions of its intended use. For a GRAS substance, generally available data and information about the use of the substance are known and accepted widely by qualified experts.

(3) Food additives whose use is governed by a regulation and described in 21 CFR 170-189. For a food additive, data and information about the use of the substance are sent by the sponsorer (*n.* 发起者) to FDA and FDA evaluate those data and information to determine whether they establish that the substance is safe under the condition of its intended use.

Japan

Ministry of Health and Welfare, currently Ministry of Health, Labor and Welfare (MHLW) in Japan enacted the Food Sanitation Act (FSA) in 1947. Under this acta positive list system for food addition was introduced. According to FSA, food additives are defined as:

(1) Substances used in or on food in the process of manufacturing food.

(2) Substances used for the purpose of processing or preserving food.

Food additives include substances remaining or not remaining in the final products. All the substances whether from natural source or synthetic for above purpose are categorized as food additives in Japan. The scope of food additives defined by FSA is not the same as defined by CAC/FDA/EU. In this act processing aids, vitamins, mineral and amino acids are also categorized as food additives.

FSA divided all food additives in following four categories:

(1) Designated additives: A positive list system for food additives last revised on January 15, 2021 and only additives designated as safe by the MHLW based on Article 10 of FSA, after risk assessment by Food Safety Commission may be used in foodstuffs. Currently, 472 additives including natural and chemically synthesized products as well as some flavorings and flavoring groups are designated as approved this regulation.

(2) Existing additives: These substances were already marketed or used on the date of the amendment of the Food Sanitation Law in 1995 and appear in the List of Existing Food Additives. This list came into force in April 1996 and now containing 365 additives.

(3) Natural flavoring agents: These substances are natural products obtained from animal and plants and used to add flavor to foods. These are generally used in very small amounts.

(4) Ordinary food used as additives: These are generally consumed foods and drinks that can be used as food additives, for example, strawberry juice for coloring, agar for stabilization.

Indian

Food additives have been mentioned in Food Safety and Stranded Act, 2006 by Indian Regulators, Food Safety and Standards Authority of India (FSSAI). The words "food additives", "ingredients" and "processing aids" have been used interchangeably in the Act. As per this Act, food additive means any substance not normally consumed as a food by itself or used as a typical ingredient of the food, whether or not it has nutritive value, the intentional addition of which to food for a technological (including organoleptic) purpose in the manufacture, processing, preparation, treatment, packing, packaging, transport or holding of such food results, or may be reasonably expected to result (directly or indirectly), in it or its by-products becoming a component of or otherwise affecting the characteristics of such food but does not include "contaminants" or substances added to food for maintaining or improving nutritional qualities. In the same act it is also mentioned that "ingredient" means any substance, including a food additive used in the manufacture or preparation of food and present in the final product, possibly in a modified form. Chapter IV of the Act also provide the provisions for the use of food additive or processing aid and indicate that no article of food shall contain any food additive or processing aid unless it is in accordance with the provisions of this Act and regulations made there under. Processing aid has been further elaborated as means any substance or material, not including apparatus or utensils, and not consumed as a food ingredient by itself, used in the processing of raw materials, foods or its ingredients to fulfill a certain technological purpose during treatment or processing and which may result in the non-intentional but unavoidable presence of residues or derivatives in the final product.

FSSAI has further regulated the use of food additives in Food Safety and Standards (Food Products Standards and Food Additives) Regulations, 2011. These regulations are dynamic and amended time to time based on the input of various stakeholders and after discussion and recommendations of a Scientific Panel on Food Additives, Flavourings, Processing aids & Materials in Contact with Food. This regulation has a specific chapter related to food additives, which includes colors and flavors apart from other food additives and this chapter also elaborates the following terms:

(1) Food in which additives may be used.

(2) Foods in which additives may not be used.

(3) ADI.

(4) Maximum Use Level.

(5) Justification for the use of food additives.

(6) GMP.

(7) Specifications for the Identity and Purity of Food Additives.

(8) Carry-over of food additives into foods.

Notes

1) There is also universal acceptance that, for a major of new food additive, adequate animal studies are necessary to address potential mutagenicity, subchronic and chronic toxicity, reproductive and developmental toxicity, and carcinogenicity at a minimum.

参考译文：人们还普遍认为，对于一种主要的新型食品添加剂有必要通过充分的动物研究，以解决其潜在的致突变性、亚慢性和慢性毒性、生殖和发育毒性以及致癌性等问题。

句中“that”引导的同位语，进一步解释说明“acceptance”。

2) Harmonization of food additive regulations, however, is an elusive goal because of major differences in global food use patterns, in the definitions of various additives and in current regulations.

参考译文：然而，由于全球食品使用模式、各种添加剂的定义和现行法规等存在重大差异，食品添加剂法规的协调统一是难以实现的目标。

句中“because of”引导的状语解释目标难以达到的原因。

Exercise

1. Translate the following sentences into Chinese.

1) The functional range of food additives is the largest in Chinese food regulation, which defines food additives as both chemosynthetic and natural substances added to food for quality, color, fragrance and taste improvement or for food preservation and processing technology.

2) The names of those food additives shall be declared in general names in accordance with GB 2760—2014. The content of each ingredient does not need to be declared. This labeling requirement is specified in GB 7718—2011.

2. Translate the following sentences into English.

1）与其他国家相比，中国将营养强化剂、口香糖中的口香糖基物质和调味剂视为直接食品添加剂。

2）食品添加剂应清楚地显示在标签的显著位置。

Reading Material 2 Flavors and Fragrances

Flavors and fragrances have a wide application in the food, feed, cosmetic, chemical, and pharmaceutical (*adj.* 制药的) sectors. Nowadays, they represent over a quarter of the world market for food additives and most of them are provided by extraction from natural sources or by traditional methods, as chemical synthesis.

The most common processes to produce flavor compounds are the extraction from natural sources and the chemical synthesis. Nevertheless, extraction from plants has many disadvantages such as low concentration of the product of interest, seasonal variation, risk of plant diseases, stability of the compound, and trade restrictions. In fact, chemical synthesis still represents the cheaper technology for their production; however, it may require harsh conditions (toxic catalysts, high pressure and temperature, among others) and usually lack adequate regio- and enantio-selectivity to the substrate, resulting in a mixture of products. Additionally, the compounds generated are labelled as "artificial" or "nature identical", decreasing their economic value. An increasing in the interest on the biotechnological production and use of flavor compounds of (micro) biological origin is observed, since the products obtained may be labeled as "natural". Even though the low yields obtained in most of the reported biotechnological processes for flavor production, in some cases they are economically feasible. Some examples of commercial "natural" flavors biotechnologically produced are ethyl butanoate (*n.* 丁酸乙酯), 2-heptanone (*n.* 2-庚酮), *β*-ionone (*n.* *β*-紫罗兰酮), nootkatone (*n.* 诺卡酮), 1-octen-3-ol, 4-undecalactone (*n.* 1-辛烯-3-醇, 4-十一内酯), and vanillin (*n.* 香草醛). One of the main motivations for the microbial production of flavor compounds is its market price, which is normally far above their synthetic counterparts (*n.* 同行), but usually lower than those extracted from nature.

Flavor compounds can be biotechnologically produced in two basic ways: through *de novo* (*n.* 从头) synthesis or by biotransformation. *De novo* synthesis refers to the production "from the new", i.e., the synthesis of substances rom simple building block molecules (sugars, amino acids, nitrogen salts, minerals, among others), which will be metabolized by organisms to form a different and complex structure. Biotransformations, in turn, are single reactions catalyzed enzymatically (as pure enzymes or within microbial cells). Therefore, the substrate is metabolized by the organism (usually a breakdown or an oxidation/reduction process) in a single (bioconversion) or a few (biotransformation) reactions to produce a structurally similar molecule. The production of aroma compounds by *de novo* synthesis usually generates a mixture of products, whose maximal concentrations are normally below 100 mg/L. Therefore, biotransformations have higher potential for the production of "bioflavors" on a commercial scale.

Biotechnological Production of Flavors

Considering the disadvantages of chemical production, regarding the quality of the product, health and environmental issues and the inability of natural production at industrial scale, the need to address an alternative way for flavor production through low-cost and environmentally friendly processes became crucial. Consumer perception that everything natural is better is causing an increase demand for natural flavor additives and biotechnological routes may be, if they exclude any chemical steps, a way to get natural products. *De novo* synthesis should be therefore used for complex targets or product mixtures, whereas bioconversions/biotransformations are able to carry out single-step processes. In general, microorganisms are able to produce a wide range of flavor compounds by *de novo* synthesis. However, the production levels are very poor, and thus constitute a limit for industrial exploitation.

Bioconversions/biotransformations can be cheaper, greener, and more direct than their chemical analogues (*n.* 类似物). Since the first discoveries of microbial production of blue cheese-note compounds in 1950, several bioflavor synthetic paths have been unveiled and exploited throughout the decades. In the next topic, the key flavor compounds of microbial production in food industry will be addressed.

Phenolic Aldehydes

The most important flavors and fragrances from the class of phenolic aldehydes(*n.* 酚醛) are anisaldehyde (*n.* 茴香醛) and some derivatives of protocatechu aldehyde (3,4-dihydroxybenzaldehyde, *n.* 原儿茶酚醛), such as vanillin, veratraldehyde (*n.* 藜芦醛), and heliotropin (*n.* 胡椒醛). In fact, vanillin is one of the most popular flavors in the world. Vanillin is the primary component of the extract of the vanilla bean (*n.* 香草豆). These flavors can be extracted from the beans of *Vanilla* species such as *V. planifolia* and *V. tahitensis*. This compound is not only widely used as flavor enhancer in sweet foods such as ice creams, cookies, or cakes, but also in soft beverages, cosmetics, or as precursor for pharmaceutical preparations, and as food preservative. Also, synthetic vanillin is used in the production of deodorants, air fresheners, cleaning products, antifoaming agents, or herbicides (*n.* 除草剂).

Terpenes (*n.* 萜烯)

Terpenoids (*n.* 萜类化合物) or isoprenoids (*n.* 类异戊二烯), are the most diversified class of natural compounds synthesized from plants, animals, or microorganisms, reaching a panoply of 40,000 different compounds. One of the most studied precursors in biotechnological monoterpenoid (*n.* 类单萜) production is limonene (*n.* 柠檬油精), since it can be derivatized to a variety of value-added compounds, such as carvone (*n.* 香芹酮), carveol (*n.* 香芹醇), perillyl alcohol (POH, *n.* 紫苏子醇), terpineols (*n.* 松果醇), menthol (*n.* 薄荷醇), and pinenes (*n.* 针烯类). Limonene (*n.* 柠烯) can be produced by one key enzyme, limonene synthase (LS), by catalysis of intramolecular cyclization of geranyl pyrophosphate (GPP, *n.* 焦磷酸香叶酯). Usually it is used as fragrance ingredient in cleaning products and perfumes or in citrus-flavored products such as candies and drinks. It has a pleasant orange or citrus-like odor and is currently produced as a side product from the citrus juice industry [citrus (*n.* 柑橘) oil can contain 70%-98% D-limonene], reaching 60,000 tons per year.

Alcohols

Alcohols are produced by the normal metabolism of the microorganisms as a result of amino acid catabolism. These compounds, such as 2-butanol (*n.* 2-丁醇), 1, 2-butanodiol (*n.* 1,2-丁二醇), and 2-phenylethanol (2-PE, *n.* 2-苯乙醇), possess (*vt.* 拥有) unique organoleptic (*adj.* 感官的) properties and are important flavor compounds in the food industry. One of the most relevant flavor-alcohols is the 2-PE, an aromatic alcohol with a delicate fragrance of rose petals (*n.* 花瓣) widely applied in diverse types of products, such as perfumes, cosmetics, pharmaceuticals, foods, and beverages. Furthermore, 2-PE can be used as raw material to produce other important flavor compounds, such as 2-phenylethylacetate (a high-value aromatic ester, *n.* 2-苯乙酸乙酯) and a potential fuel molecule, phenylacetaldehyde (*n.* 苯乙醛), and p-hydroxphenylethanol (*n.* 对羟基苯乙醇), used in pharmaceutical and fine chemical industries.

Lactones (*n.* 内酯)

Lactones are cyclic esters derived from lactic acid and they are constituents of a wide variety of essential oils and plant volatiles. They are well-known for their great variety of taste and aroma (peach, pineapple, apricot, raspberry, strawberry, mango, papaya, cream, coconut, and nutt-like) (桃子、菠萝、杏子、覆盆子、草莓、芒果、木瓜、奶油、椰子和坚果类), reaching a production of hundreds of tons per year. The most important lactones are five- and six-membered rings, γ-and δ-lactones, respectively, with equal or less 12 carbons. It includes compounds such as 4-dodecanolide (*n.* 4-十二烷醇, coconut-fruity like), 4-decanolide/ γ-decalactone (*n.* 4-癸醇, peach-like), 4-octanolide (*n.* 4-辛醇, sweet hearbaceous coconut-like),

5-dodecanolide (*n.* 5-十二醇，fruit-oily peach-like), 5-decanolide/δ-decalactone (*n.* 5-癸醇，creamy-coconut peach-like), and 6-pentyl-α-pyrone (6PP, *n.* 6-戊基-α-吡喃酮，strong coconut-like). Currently, lactone production is mainly achieved by chemical synthesis from keto acids; however, biotechnological production of γ-decalactones (GDL, *n.* γ-十内酯) and δ-decalactones (DDL, *n.* δ-十内酯) is growing due to their natural/GRAS label.

De Novo Synthesis

Flavor and fragrance production by *de novo* synthesis uses the whole metabolism of the microorganism to produce a combination of flavor compounds, in contrast to the biotransformation, where a specific reaction(s) produce a major compound. Whole cells catabolize carbohydrates, fats, and proteins, and further convert the breakdown products into flavor components. Nevertheless, under these conditions, only trace amounts of flavors (more complex compounds) are produced and this process is not very promising and economically viable for industrial production because of the low concentrations of produced flavors. Some of the most relevant flavor compounds obtained by means of *de novo* synthesis is showed in below (Table 4.3).

Table 4.3 Flavor compounds obtained from microbial *de novo* synthesis

Product	Aroma description	Host	Concentration (g/L)	Process information
2-Phenylethanol	Rose-like	*Yarrowia lipolytica*	0.2	Shake flask
		Kluyveromyces marxianus	1.3	Shake flask
		Escherichia coli	0.26	Shake flask
		K. lactis	0.72	Shake flask
Diacetyl	Buttery	*Lactobacillus casei*	1.4	MRS medium
		L. lactis	0.36	Skim milk medium
		Enterobacter aerogenes	1.35	CMD medium
		Candida glabrata	4.7	CMD medium
Limonene	Orange-like	*E. coli*	430 mg・L^{-1}	Dodecane organic phase
Vanilin	Vanilla	*E. coli*	119 mg・L^{-1}	From glucose
		Saccharomyces cerevisiae	500 mg・L^{-1}	From glucose

Notes

1) Considering the disadvantages of chemical production, regarding the quality of the product, health and environmental issues and the inability of natural production at industrial scale, the need to address an alternative way for flavor production through low-cost and environmentally friendly processes became crucial.

参考译文：考虑到化学品生产在产品质量、健康和环境以及无法在工业规模上进行自然生产等缺点，如何通过低成本和环境友好的工艺来解决香料生产问题变得至关重要。

句中“Considering”引导的状语表示基于…考虑，“the need became crucial”是句子的主干，“to”引导的短语修饰“need”。

2) Consumer perception that everything natural is better is causing an increase demand for natural flavor additives and biotechnological routes may be, if they exclude any chemical steps, a way to get natural products.

参考译文：消费者认为一切天然的更好，这导致对天然香料添加剂的需求增加。如果不包括任何化学步骤，生物技术方法是获得天然产品的一种方式。

句中“that”引导的同位语进一步解释“Consumer perception”，“if”引导的条件状语表示“biotechnological routes may be way to get natural products”成立的条件。

Exercise

1. Translate the following sentences into Chinese.

1) *De novo* synthesis should be therefore used for complex targets or product mixtures, whereas bioconversions/biotransformations are able to carry out single-step processes.

2) Lactones are cyclic esters derived from lactic acid and they are constituents of a wide variety of essential oils and plant volatiles.

2. True or false.

1) All flavor compounds used in food are the extraction from natural sources. (　　)

2) Flavor compounds can be biotechnologically produced in two basic ways: through *de novo* synthesis or by biotransformation. (　　)

Reading Material 3　Processing Aids

Processing aids are used by manufacturers to help solve many product-processing needs without being required to be declared on the food label by law in most countries. Processing aids cover everything from the lubricants (*n.* 润滑剂) used on equipment that come into contact with food to agents that help with peeling fruits and vegetables. They are used to provide many useful functions and desired effects during the manufacture of foods but are not meant to be part of the final product's ingredients. Thus, they are not required to appear on a product's ingredient label.

According to the *Codex Alimentarius Standard* (1981), processing aids means any substance or material, not including apparatus(*n.* 仪器) or utensils(*n.* 用具), and not consumed as a food ingredient by itself, intentionally used in the processing of raw materials, foods or its ingredients, to fulfil a certain technological purpose during treatment or processing and which may result in the non-intentional but unavoidable presence of residues or derivatives in the final product.

(1) Substances that are added during the processing of a food but are removed in some manner from the food before it is packaged in its finished form.

(2) Substances that are added to a food during processing that are converted into constituents normally present in the food, and do not significantly increase the amount of the constituents naturally found in the food.

(3) Substances that are added to a food for their technical or functional effect during processing but are present in the finished food at insignificant levels and do not have any technical or functional effect in that food.

An incomplete list of process aids and their purposes is given in Table 4.4.

Table 4.4 Examples and functions of processing aids

Food Products	Processing Aid	Purpose
Apple juice	Gelatin with gums	Helps to eliminate suspended particles
Baked goods and baking mixes	Agar	Vegan substitute for gelatin that helps the gelling of mixes
Beverages	Silicone	Antifoam
Bread	Phospholipase	Increase volume and prolongs softness
Cheese	Rennet	Separates curd and whey
Chill water	Ozone	Antimicrobial
Dough	Xylanase	Increases flexibility
Fish and meat (seafood)	Salt	Decrease water activity to improve shelf life
Frozen dough (e.g., waffles and pancakes)	Sodium sterol lactylate	Strengthens dough
Fruit and vegetable washes	Chlorine organic acid washes	Antimicrobial
Liquid nitrogen	BBQ sauce	Improves stability of plastic container
Meat	Ammonium hydroxide	Antimicrobial
Products transported on conveyors	Oil or synthetic	Lubricant
Sugar	Dimethylamine epichlorohydrin copolymer	Decoloring agent helps in clarification of sugar

No generalized regulatory criteria exist in most countries for judging exactly what constitutes an insignificant level of a processing aid. Each application to be considered as a processing aid must be submitted to the relevant regulatory authority where such an authority exists and where it takes on the responsibility of regulating process aids. This will usually require that the decision for a particular compound for a particular application will need to be handled individually. This leads to further confusion in the marketplace and with consumers because these decisions are being worked out differently in different countries using a process that is not always public and to date most national authorities have not tried to regulate the use of processing aids in other countries.

Owing to the advantages of using processing aids, they are not likely to be eliminated. However, continuous improvements in processing methods and equipment may make them obsolete (*adj.* 淘汰). Improvements in the formulation and application of processing aids also might make them more effective and more ethical for their intended use. Finally, companies may opt to select more ethical processing aids or include them in their ingredient list. Processing aids are not required to be listed on the label, but some trace amounts of the material may remain in the product. Also, some processing aids are converted to normal constituents of the food but must not significantly increase the original amount. In any case, a processing aid is required to be GRAS. This means the overwhelming evidence considered by industry, academia and independent experts agrees the processing aid is safe for consumers.

Notes

1) According to the *Codex Alimentarius Standard* (1981), processing aids means any substance or material, not including apparatus or utensils, and not consumed as a food ingredient by itself, intentionally used in the processing of raw materials, foods or its ingredients, to fulfil a certain technological purpose during treatment or processing and which may result in the non-intentional but unavoidable presence of residues or derivatives in the final product.
参考译文：根据《食品法典委员会标准》（1981 年），加工助剂是指为了在处理或加工过程中达到一定的技术目的而在加工原材料、食品或其成分时故意使用的任何物质或材料，可能导致最终产品中非故意但不可避免地存在残留物或衍生物，其中不包括仪器或器具。
句中“...to”引导的动词不定式修饰“used”表示目的，“which”引导定语从句。
2) This leads to further confusion in the marketplace and with consumers because these decisions are being worked out differently in different countries using a process that is not always public and to date most national authorities have not tried to regulate the use of processing aids in other countries.
参考译文：这导致市场和消费者进一步混淆，因为这些决定在不同的国家以不同的方式制定，使用的过程并不总是公开的，而且迄今为止，大多数国家当局还没有试图监管其他国家加工助剂的使用。
句中引起“confusion”的原因包括两方面，之间用“and”连接。

Exercise

1. Translate the following sentences into Chinese.

1) They are used to provide many useful functions and desired effects during the manufacture of foods but are not meant to be part of the final product's ingredients.

2) Each application to be considered as a processing aid must be submitted to the relevant regulatory authority where such an authority exists and where it takes on the responsibility of regulating process aids.

2. True or false.

1) Processing aids are one of the ingredients of food and must be indicated in the label. ( )

2) Some substances are both food additives and processing aids. ( )

Chapter 5

Food Processing

Lesson 1 Dairy Product Processing

Milk is defined as the secretion of the mammary glands of mammals, its primary nature function being the nutrition of young.

Principal Components

A classification of the principal constituents of milk is given in Table 5.1. The principal chemical components or groups of chemical components are those present in the largest quantities. Of course, the quantity (in grams) is not paramount in all respects. For example, vitamins are important with respect to nutrition value; enzymes are catalysts of reactions; and some minor components contribute markedly to the taste of milk.

Table 5.1 Approximate Composition of Milk

Component	Average content in milk (% *W/W*)	Range [a] (%*W/W*)	Average content in dry matter (%*W/W*)
Water	87.1	85.3-88.7	—
Solids-not-fat	8.9	7.9-10.0	—
Lactose	4.6	3.8-5.3	36
Fat	4	2.5-5.5	31
Protein [b]	3.3	2.3-4.4	25
Casein	2.6	1.7-3.5	20
Mineral substances	0.7	0.57-0.83	5.4
Organic acids	0.17	0.12-0.21	1.3
Miscellaneous	0.15	—	1.2

Note: Typical for milks of lowland breeds.

a. These values will rarely be exceeded, e.g., in 1%-2% of samples of separate milkings of healthy individual cows, excluding colostrum and milk drawn shortly before parturition.

b. Nonprotein nitrogen compounds not include.

Milk Preservation Methods

The manufacture of milk products virtually always involves some forms of preservation, which means taking measures to prevent or at least postpone, deterioration. Most technologists primarily think of deterioration caused by microorganisms, but it can also involve enzymatic, chemical, or physical changes.

To counteract microbial action, one can (a) kill the microorganisms, (b) physically remove them, (c) inhibit their growth (although this will not always prevent metabolic action by the enzyme systems of the organism), and (d) prevent contamination with microbes. To counteract enzymatic action one can irreversibly inactivate the enzyme; the resistance substantially varies among enzymes and with environmental conditions. One can also reduce the specific activity of an enzyme by changing the environment.

Several preservation methods can be applied, and all have specific advantages and disadvantages. We will briefly discuss the more important methods.

Heat treatment. This is generally the method of choice for liquid products. It is active against microbes and enzymes. The method is convenient, flexible, well-studied, and fairly inexpensive. The disadvantage is that undesirable chemical reactions occur, especially at high heating intensity, for instance, causing off-flavors.

Pressure treatment. The hydrostatic pressure applied must be high, well over 100 MPa. The high pressure leads to unfolding of globular proteins and thereby killing of microbes and inactivation of some enzymes. For example, to reduce the number of vegetative bacteria by a factor of 10^5 or 10^6, a pressure of about 250 MPa must be applied for 20 min or 500 MPa for 10 s. Spores, as well as most enzymes, are far more resistant. The great advantage of this method is that undesirable chemical reactions hardly occur. A disadvantage is that in milk, the casein micelles tend to dissociate irreversibly, leading to a significantly changed product. Moreover, the method is expensive because the process is discontinuous and can only be applied to relatively small volumes at a time. It is not applied for milk products.

Irradiation. This can be ionizing radiation, for example, β- or γ-rays emitted by radioactive materials, or ultraviolet light. The former needs to be of high intensity to kill bacteria, especially spores. This causes off-flavors. Moreover, there is considerable public opposition against the use of radioactive materials. The method is only used for some condiments and for sterilizing surfaces. UV light kills microbes, but it can only penetrate clear liquids. It is occasionally used for water sterilization and also for surface decontamination. Neither of these methods is suitable for inactivation of enzymes.

PEF, i.e., short pulses of a high electric field. Such pulses can kill microbes, presumably, by damaging the cell membrane. The smaller the cell dimensions, the higher the intensities needed, and spores are very difficult to kill. Enzymes are generally not inactivated. The method is not used for milk products.

Removal of microbes. This can have the obvious advantage that no chemical reactions occur. On the other hand, enzymes are not inactivated, and the complete removal of microbes cannot generally be achieved. At high concentrations of water-soluble substances, most microorganisms stop growing, presumably because the contents of the cell become highly concentrated. There is considerable variation among microbes and with the nature of the solute, but a concentration corresponding to a water activity below 0.65 suffices in most milk products. Dried milk is thus free from microbial growth. To stop enzyme action, lower water activities are generally needed, below 0.2 or even less.

Freezing. This causes freeze concentration and acts much like evaporation or drying; moreover, the temperature is so low that microbial or enzymatic action is anyway very sluggish.

Mild preservatives. This means high concentrations of salt (e.g., cheese), acid (e.g., fermented milk), or sugar (e.g., sweetened condensed milk). Acids act in their undissociated form, and thus need a pH low enough to greatly decrease their dissociation to be effective. Acids and salts can substantially decrease enzyme action; sugars generally do not. The preservatives naturally affect product properties, especially flavor. Disinfectants should never be added to milk or milk products.

Milk Storage and Transportation

Milk storage and transport operations are aimed at having good-quality milk available where and when needed for processing. The milk should not be contaminated by microorganisms, chemicals, water, or any other substance. Obviously, the costs involved in storage and transport should be kept low, which implies that, for example, loss of milk should be minimized. Simple and effective cleaning of all the equipment involved should be possible.

Liquid Milk Processing

Liquid milk can be delivered to the consumer after various heat treatments: none (raw milk), pasteurized or sterilized, and either packaged or not (although sterilized milk is, of course, always packaged). The properties of liquid milk that require the most attention are safety to the consumer, shelf life, and flavor. Safety is, of course, essential and consumption of raw milk cannot be considered safe. Consequently, the delivery of raw milk is prohibited or severely curtailed in many countries. Likewise, delivering milk that is not packaged may involve health hazards.

Pasteurized Milk

Pasteurized beverage milk must be safe for the consumer and have a shelf life of a week or longer when kept refrigerated. Flavor, nutritive value, and other properties should deviate only slightly from those of fresh raw milk.

Thermalization can prevent fat and protein breakdown by heat-resistant enzymes of psychrotrophic bacteria. The keeping time of pasteurized milk is too short to cause noticeable decompositions by these enzymes, unless the original milk had a high count of psychrotrophs.

Separation is needed to adjust to the desired fat content. If homogenization is omitted, only a part of the milk will be skimmed, while the skim milk volume obtained should suffice to standardize the milk.

Homogenization serves to prevent the formation of a cream layer in the package during storage. In low-pasteurized milk, a loose cream layer of agglutinated fat globules forms that can be easily re-dispersed throughout the milk. In high-pasteurized milk, the cold agglutinin has been inactivated and a cream layer forms far more slowly, but then it is a compact, hardly dispersible layer; a solid cream plug may even result from partial coalescence of the fat globules. Therefore, this milk is usually homogenized.

ESL Milk

Some consumers desire beverage milk that tastes like low-pasteurized milk, but that can be kept much longer without perceptible quality loss. There are two principles by which such ESL milk can be produced.

The first involves UHT-heat treatment, followed by aseptic packaging. This actually results in sterilized milk. However, a heat treatment of 140℃/2 s or 135℃/3 s will suffice to kill all bacteria, while it can leave the flavor virtually unaltered, provided that direct heating is applied. The milk must be free of enzymes produced by psychrotrophs, as these are not inactivated.

Another possibility is removal of microbes by microfiltration, which has met with some success. The transmembrane pressures applied are below 100 kPa. A high flux and long operating periods can be achieved. The fat globules are also retained, considering that the membrane has a pore size of about 1 μm; therefore, the milk should first be separated. Some 0.1%-1% of the total number of bacterial cells passes to the permeate, of *B.cereus* <0.05%. Stronger reductions, even up to sterility, can be obtained by using membranes with smaller pore sizes, but that is at the expense of the flux and of the maximum operating time.

UHT-sterilized Milk

Sterilization of milk is aimed at killing all microorganisms present, including bacterial spores, so that the packaged product can be stored for a long period at ambient temperature, without spoilage by microorganisms. Since molds and yeasts are readily killed, we are only concerned about bacteria. The undesirable secondary effects of in-bottle sterilization like browning, sterilization flavor, and losses of vitamins can be diminished by UHT sterilization. After UHT sterilization certain enzymatic reactions and physicochemical changes still may occur.

The proteinases and lipases of psychrotrophs, especially of the genus pseudomonas, can be very heat resistant and even in-bottle sterilization does not suffice to fully inactivate these enzymes. Therefore, the enzymes should be absent in the raw milk. In particular, the addition of some milk left over for some time should be carefully avoided because in this milk psychrotrophs may have grown extensively.

UHT sterilization is mostly performed at temperatures above 140℃. Accordingly, the sterilizing effect required is readily attained. But a sufficiently long shelf life at ambient temperature is only obtained if the residual activity of plasmin is at most 1%. Often, the curve for 600 mg lactulose represents the upper limit of UHT sterilization, but at that limit significant cooked flavor results.

Reconstituted Milk

In several regions, there is a shortage of fresh (cows') milk. As an alternative, milk powder can be used to make a variety of liquid milk products. Some common types are the following:

Reconstituted milk. It is simply made by dissolving whole milk powder in water to obtain a liquid that is similar in composition to whole milk. Likewise, reconstituted skim milk can be made.

Recombined milk. It is made by dissolving skim milk powder in water, generally at 40-50℃, then adding liquid milk fat (preferably anhydrous milk fat of good quality), making a coarse emulsion by vigorous stirring or with a static mixer, and then homogenizing the liquid. This product is similar to homogenized whole milk, except that it lacks most of the material of the natural fat globule membrane, such as phospholipids.

New Words

micelle [maɪ'sel] *n.* 胶束；胶团；胶粒

dissociate [dɪ'səʊʃieɪt] *vt.* 分离；解离；使脱离关系

irreversibly [ˌɪrɪ'vɜːsəbli] *adv.* 不可逆地

deterioration [dɪˌtɪərɪə'reɪʃn] *n.* 变质；变坏

disinfectant [ˌdɪsɪn'fektənt] *n.* 消毒剂；杀菌剂；消毒水
adj. 消毒的

coalescence [ˌkəʊə'lɛsns] *n.* 合并；联合；接合

skimmed [skɪmd] *vt.* 撇去乳脂；脱脂

sterilization [ˌsterəlaɪ'zeɪʃn] *n.* 灭菌；消毒

plasmin ['plæzmɪn] *n.* 血纤维蛋白溶酶；血浆酶；胞浆素

psychrotrophs [saɪkrəʊt'rɒfs] *n.* 耐冷菌；嗜冷菌

pseudomonas [(p)sju'dɒmənəs] *n.* ［微］假单胞菌

lactulose [ˌlæktʊ'ləʊz] *n.* 半乳糖苷果糖；乳果糖

reconstituted [ri'kɒnstɪtjuːtɪd] *adj.* 再造的；再生的

emulsion [ɪ'mʌlʃn] *n.* 乳状液；乳浊液；感光乳剂

agglutinate [ə'gluːtɪneɪt] *adj.* 黏着的；胶合的

v. 使…黏着；成胶状；黏合

Notes

1) Homogenization serves to prevent the formation of a cream layer in the package during storage. In low-pasteurized milk, a loose cream layer of agglutinated fat globules forms that can be easily re-dispersed throughout the milk.

参考译文：均质化的作用是防止在储存期间包装中形成奶油层。在低度巴氏杀菌的牛奶中，形成了由凝集的脂肪球组成的松散奶油层，可以很容易地重新分散到整个牛奶中。

2) Sterilization of milk is aimed at killing all microorganisms present, including bacterial spores, so that the packaged product can be stored for a long period at ambient temperature, without spoilage by microorganisms.

参考译文：牛乳的灭菌旨在杀灭包括细菌孢子在内的所有微生物，从而保证包装食品能在室温下长时间保存而不会出现微生物腐败。

Lesson 2 Meat Product Processing

Processed meat products are defined as those in which the properties of fresh meat have been modified by the use of one or more procedures, such as grinding or chopping, the addition of seasonings, alternation of colour, or heat treatment. Most of the meat products are subject to a combination of several basic processing steps before reaching their final form. Although each processed product has its own specific characteristics and methods of preparation, they all can be classified as either comminuted or noncomminuted products.

Historical Background

Meat product processing originated in prehistoric times, and no doubt developed soon after people became hunters. Probably the first type of processed product was sun dried meat, and only later was meat dried over a slow burning wood fire to give a dried, smoked meat similar to jerky. The salting and smoking of meat was an ancient practice even in the time of Homer, 850 B.C. These early processed meat products were prepared for one purpose, their preservation for use at some future time. People had learned at a very early time that dead or heavily salted meat would not spoil as easily as the fresh product. Meat processing probably developed out of this knowledge, coupled with the necessity for storing meat for future use. With advances in preservation technology, especially in refrigeration and packaging, meat processors were no longer tied primarily to preservation by high salt concentration or drying. They were free to experiment with lower salt and higher moisture levels in the finished product, and with new seasonings and combinations of meat ingredients, thereby creating many new processed products. Reasons for preparation of modern processed meat products include development of unique flavors and forms of product, provision of a variety of products, and development of new products in addition to preservation of meat.

Basic Processing Procedures

Meat processing technology comprises the steps and procedures in the manufacture of processed meat products. Processed meat products, which include various different types and local/regional variations, are food of animal origin, which contribute valuable animal proteins to human diets. Animal tissues, in the first place muscle meat and fat, are the main ingredients, besides occasionally used other tissues such as internal organs, skins and blood or ingredients of plant origin. All processed meat products have been in one way or another physically and/or chemically treated. These treatments go beyond the simple cutting of meat into meat cuts or meat pieces with subsequent cooking for meat dishes in order to make the meat palatable. Meat processing involves a wide range of physical and chemical treatment methods, normally combining avariety of methods. Meat processing technologies include: cutting/chopping/comminuting, mixing/tumbling, salting/curing, utilization of spices/non-meat additives, stuffing/filling into casings or other containers, fermentation and drying, heat treatment and smoking.

Curing

The two main curing ingredients that must be used to cure meat are salt and nitrite. However, other substances can be added to accelerate curing, stabilize color, modify flavor, and reduce shrinkage during processing.

Salt is the primary ingredient used in meat curing. Originally it served as a preservative by dehydration and osmotic pressure which inhibits bacterial growth. The main function of salt in other cured products is to add flavor. However, even at low concentrations salt has some preservative action. Salt levels are dependent on consumers' taste, but a 2%-3% concentration in the product is about right.

Nitrates and nitrites, either potassium or sodium salt, are used to develop cured meat color. They impart a bright reddish, pink color, which is desirable in a cured product. In addition to the color role, nitrates and nitrites have a pronounced effect on flavor. Without them a cured ham would be simply a salty pork roast. They further affect flavor by acting as powerful antioxidants.

Antioxidants are compounds that prevent the development of oxidative rancidity, which would reduce the keeping quality. Sodium nitrites also prevent the growth of a food poisoning microorganism known as *Clostridium botulinum*, the bacteria that causes botulism.

Sugar (sucrose) serves several important purposes in cured meat. First of all, it adds flavor, and secondly, it counteracts the harshness of salt. Also, sugar provides a surface color characteristic of aged ham if caramelized sugar is used. Both brown and white sugars can be used. The sugars most frequently used are sucrose, cane sugar, dextrose, and invert sugar. The amount of sugar used in self-limiting due to its sweetening power.

Ascorbates (sodium ascorbate or sodium erythorbate) are used to speed the curing reaction by faster color development through more rapid reduction of nitrates and nitrites to nitric oxide. The nitric oxide combines with the meat pigment, myoglobin, to form nitrosomyglobin, dark red color. When the product is heated to 55-60℃, the nitrosomyglobin is converted to a stable pigment, nitrosohemochrome, light pink in color. Also, ascorbates are used to stabilize the cure color of meats.

Curing materials may be in either dry or liquid form. They will be applied either to the surface of meat or into it by some injection method. The oldest method of cure application is dry cure in which the curing ingredients are rubbed on the surface of the meat. The dry cure method can be used under wider temperature variations and will have less spoilage problems under unfavorable curing conditions.

During curing the product should be stored at 0-4℃. The length of curing is 7 d per inch of thickness. A belly two-inches thick should cure in 14 d. Curing time for hams and picnics in brine is 3.5-4 d per pound per piece of meat.

To speed up curing, the brine can be pumped or injected into the cut. Injection of the cure is accomplished either by stitch or artery method. The stitch method involves the use of a perforated needle or several needles that distribute the pickle when injected into the meat.

Another type of cure injection used especially for hams, picnics and beef tongues is by the arterial system. This procedure utilizes the naturally occurring vascular network for quicker and complete distribution of the cure. The pickle is pumped by a small gauge needle through the femoral artery of the ham. Any type of injection curing will speed up the distribution, and the more complete the distribution, the shorter the curing time. During the curing period, the product should be kept at 2-4℃.

Smoking

The smoking of meat is the process of exposing a product to wood smoke at some point during its manufacture. Smoking methods originated simply as a result of meat being dried over wood fires. In most present day processed meats, smoking contributes little if any preservative action. Smoke components are absorbed by the surface and interstitial water in the product, but in no case do they penetrate more than a few millimeters. In products where the surface remains intact, a preservative effect will persist. However, a few other advantages do accrue from the smoking of meats. For example, it aids in the development of a smooth surface or skin beneath the cellulose casing of frankfurters that facilitates peeling of the casing prior to packaging.

Three traditionally recognized reasons for smoking meat are for preservation, appearance, and flavor. Smoked meat is less likely to spoil than unsmoked meat. This is due to the bactericidal and bacteriostatic properties of smoke. These properties are attributable to certain components in the smoke, such as phenols and acids. Smoking improves the flavor and appearance, aids in reducing mold growth as well as retards rancid flavors. It takes about 24 h to smoke and cook hams.

Dehydration

Dehydration of meat products is one of the several basic processing steps. However, few meat products are dehydrated as a separate process. In those cases where drying is a separate step, the objective is primarily preservation. Drying to preserve the product can be accomplished by freeze dehydration or by the application of heat.

Drying meat under natural temperatures, humidity and circulation of the air including direct influence of sun rays, is the oldest method of meat preservation. It consists of a gradual dehydration of pieces of meat cut to a specific uniform shape that permits the equal and simultaneous drying of whole batches of meat.

As a general rule, only lean meat is suitable for drying. Visible fatty tissues adhering to muscle tissue have a detrimental effect on the quality of the final product. Under processing and storage conditions for dry meat, rancidity quickly develops, resulting in flavour deterioration. Dry meat is generally manufactured from bovine meat although meat from camel, sheep, goats and venison (e.g., antilopes, deer) is also used. The meat best suited for drying is the meat of a medium-aged animal, in good condition, but not fat. Meat from animals in less good nutritional condition can also be used for drying, but the higher amount of connective tissue is likely to increase toughness.

Typical Meat Products

(1) Sausages: The term sausage is derived from the Latin word "*salsus*" meaning salt, or, literally translated, refers to chopped or minced meat preserved by salting. Sausages are one of the oldest forms of processed foods, their origin being lost in antiquity. It has been reported that sausages were used by the Babylonians and the Chinese about 1500 B.C, although documented proof for this is lacking. Usually, sausage is made from ground meat with a skin around it. Typically, it is formed in a casing traditionally made from intestine, but sometimes synthetic. Some sausages are cooked during processing and the casing may be removed after. Sausage

making is a traditional food preservation technique and they may be preserved by curing, drying, or smoking.

(2) Salami: It is cured sausage, fermented and air-dried meat, originating from one or a variety of animals. Historically, salami was popular among Southern European peasants because it can be stored at room temperature for periods of up to 30-40 d once cut, supplementing a possibly meager or inconsistent supply of fresh meat. Varieties of salami are traditionally made across Europe.

(3) Ham: Ham is a processed pork foodstuff, which undergoes preservation through curing, smoking, or salting. Ham was traditionally made only from the hind leg of swine, and referred to that specific cut of pork. Ham is made around the world, including a number of regional specialties, although the term now has wider usage and can also be used to refer to meat which has been through a process of re-forming. The precise nature of meat termed “ham” is controlled, often by statute, in a number of areas, including the US and EU, meaning that only products meeting a certain set of criteria can be called ham. In addition, there are numerous ham products which have specific geographical naming protection, such as Prosciutto di Parma in Europe, and Smithfield ham in the US. Ham is also often colloquially used to mean any sliced (or re-formed) preserved meat, regardless of the animal it is made from, although this is usually qualified with the species of animal as with turkey ham.

Summary

Successful processing is dependent on proper handling of meat and using good quality ingredients. Nowadays, there is a tendency to use milder preservation methods, either because of energy-saving, because of the consumers’ preference for mildly cured or cooked products, or because of an aversion to the preservatives. In order to satisfy these demands, it will be necessary for the meat industry to improve the processing methods ceaselessly.

New Words

grinding ['graɪndɪŋ]	*adj.*	磨的；碾的；摩擦的 动词 grind 的现在分词形式
chopping ['tʃɒpɪŋ]	*adj.*	切碎的；剁碎的 动词 chop 的现在分词形式
salting ['sɔːltɪŋ]	*n.*	盐渍
curing ['kjʊərɪŋ]	*n.*	腌制
nitrite ['naɪtraɪt]	*n.*	亚硝酸盐
shrinkage ['ʃrɪŋkɪdʒ]	*n.*	收缩；减少；损失

dehydration [ˌdiːhaɪ'dreɪʃn]	*n.*	脱水
osmotic [ɒz'mɒtɪk]	*adj.*	渗透性；渗透的
rancidity [ræn'sɪdɪtɪ]	*n.*	腐败；恶臭；腐臭气味
Clostridium botulinum	*n.*	肉毒杆菌
caramelized ['kærəməlaɪzd]	*vt.*	（糖）熔化而变成焦糖
ascorbate ['æskɔːbeɪt]	*n.*	抗坏血酸盐
erythorbate [ˌerɪ'θɔːbeɪt]	*n.*	异抗坏血酸盐
myoglobin ['maɪəˌgləʊbɪn]	*n.*	肌球素；肌红蛋白
nitrosomyoglobin [naɪtrsəma'ɪəʊgləʊbɪn]	*n.*	亚硝基肌红蛋白
synthetic [sɪn'θetɪk]	*adj.*	合成的；人造的；综合的；虚伪的
	n.	合成物；人工制品
salami [sə'lɑːmi]	*n.*	意大利香肠
ham [hæm]	*n.*	火腿；火腿肉

Notes

The nitric oxide combines with the meat pigment, myoglobin, to form nitrosomyglobin, dark red color. When the product is heated to 55-60℃, the nitrosomyglobin is converted to a stable pigment, nitrosohemochrome, light pink in color.

参考译文：一氧化氮与肉中的色素肌红蛋白结合，形成亚硝基肌红蛋白，是暗红色。当产品被加热到55～60℃时，亚硝基肌红蛋白转化为一种稳定的色素亚硝基血色原，为浅粉红色。

Lesson 3 Fruit and Vegetable Processing

Fruit and vegetables contain a diverse range of bioactive constituents that may be beneficial to human health, including dietary fibers, vitamins, minerals, carotenoids, and polyphenols. Eating a diet rich in fruits and vegetables may improve health and wellbeing by reducing the risk of diseases, such as cardiovascular diseases and cancer. Moreover, consuming more plant-based foods (rather than animal-based ones) also has benefits to the health of our planet, by reducing greenhouse gas emissions and other forms of pollution, decreasing land and water use, and increasing biodiversity. The growing awareness of the health and environmental benefits of plant-based foods has led to increasing consumer demand for these kinds of product, which in turn, has prompted food suppliers to develop more plant-based products.

Processing Methods

Thermal Processing

Thermal processing is the most commonly used approach for the manufacture and preservation of fruit- and vegetable-based products. It includes a diverse range of methods (60-200℃), such as pasteurization, sterilization, steaming, cooking, boiling, roasting, and microwaving, which can be applied to either solid or liquid foods.

Non-thermal Processing

Based on the growing demand for more minimally processed fruit and vegetable-based products, many non-thermal processing technologies have been investigated, since these innovative technologies are often less detrimental to food quality attributes than thermal processing. A possible reason is that non-thermal processing generally affects noncovalent bonds of molecules, while thermal processing destroys noncovalent bonds and covalent concurrently. Non-thermal processing technologies are among those with increased interest in the sustainable production. For example, HPH is more environmentally sustainable than traditional thermal processing treatments due to lower resources consumption (water and energy). In recent years, most of the studies regarding the effects of non-thermal processing on the structure and functionality of pectin in foods have focused on HHP, HPH and US treatments, a few on PEF, HPCD and others.

(1) High hydrostatic pressure (HHP)

HHP transmits the pressure using a fluid (generally water) as the medium and the pressure (100-1,000 MPa) is applied isostatically from all directions in a confined space. Indeed, HHP predominantly affects noncovalent bonds (such as hydrogen bonds, ionic bonds, and hydrophobic bonds) and induce structural changes of biological macromolecules, such as dietary fiber (e.g., pectin) and protein (pectin-related enzymes), which affects their functional properties. During HHP processing, different pressure and temperature combinations can be used to achieve desired effects on fruit- and vegetable-based products.

(2) High pressure homogenization (HPH)

Conventional HPH is normally conducted up to pressures of 200 MPa, whereas ultra-high pressure homogenization (UHPH) is conducted at higher pressures (350-400 MPa). Generally, HPH is carried out by forcing a fluid through a small orifice using a piston. The rapid change in fluid velocity and pressure when it passes through the valve generates turbulence, shear and cavitation forces that can disrupted polymers, particles or cells in the system. Homogenization could generate changes in the structure of the enzyme via modifying mainly weak bonds, thus causing different denaturation levels and, consequently, affecting the catalytic activity of enzymes. The important processing parameters for HPH treatment are the valve design, homogenization pressure, number of passes, and operating temperature.

(3) Ultrasound (US)

US treatment is based on the use of powerful high frequency pressure waves that are too high pitched for the human ear to detect (typically 20-100 kHz).US can be directly applied to liquid food products, but solid food products have to be processed using certain fluids as a transmitting medium. The effects of US treatment are primarily ascribed to the acoustic cavitation phenomenon, which is caused by microbubble generation, growth, and implosion during the propagation of the high-intensity ultrasonic wave. Based on acoustic cavitation, US has been widely applied in oriented modification and degradation of pectin. The main parameters of US treatment involve intensity, frequency, sonication time, solvent and temperature.US could also cause the breakdown of the hydrogen bonding and Van der Waals interactions in the polypeptide chains, leading to modifications of the secondary and tertiary structure of the protein.

(4) Pulsed electric field (PEF)

PEF is conducted by putting food products in a chamber containing two electrodes and then subjecting them to short-duration high-voltage pulses, which causes membrane electroporation. The main parameters for PEF are electric field intensity, pulse frequency, pulse duration, and number of pulses. Previous research has shown that these parameters can be tuned to modify the structure of various kinds of macromolecules.

(5) High pressure carbon dioxide (HPCD)

In HPCD processing, products are contacted with either sub- or super-critical CO_2 for a fixed period. Supercritical CO_2 is carbon dioxide that is held at a temperature and pressure above its critical point (T_c=31.1℃, P_c=7.38 MPa), which causes it to exist as a single phase where it can diffuse through solids like gas and dissolve materials like a liquid. HPCD processing involves dissolving gasses in food materials by applying high pressures, followed by rapid decompression to atmospheric pressure, which results in disruption of the materials.

The Influence of Fruit and Vegetable Processing on the Physica and Chemical Properties of Pomace

Food processing involves a wide variety of processing methods and parameters, but usually results in a lower quality of products in terms of reduction in the content of polyphenols (mainly anthocyanins) in fruit and vegetables. On the other hand, processing extends the shelf-life of food products. The progress in food processing technologies allows retaining ever higher levels of bioactives in foods by combining different processing methods or selecting appropriate processing parameters. Fruit or vegetable pomace can be used as a natural food additive to enrich/supplement foods with bioactive compounds. But to make sure that the quality of an additive approximates that of the input material, processing must be carefully designed, as pomace preparation may significantly modify its composition and thus affect its quality already at the start of processing.

One of the best-known food preservation methods is drying, which with an appropriate parameter selection may minimize losses of bioactive compounds. For example, techniques selected for drying blackcurrant pomace were found to significantly affect the physical and chemical properties of obtained powders. Vacuum-microwave drying resulted in greater moisture reduction than convective and combined drying (convective pre-drying and microwave vacuum finish drying), which was attributable to considerably higher temperature applied during vacuum microwave drying as compared with convective drying (over 100℃). It was found out that the higher the temperature was during convective drying, the lower was the solubility of the obtained powders, with the control (freeze-dried) sample being the best soluble of all processed samples. Drying changes the product's chemical properties. In numerous studies on blackcurrant pomace processing anthocyanins were observed to be the dominant group of polyphenols present in blackcurrant pomace powders. The lowest degradation of polyphenolic compounds was noted after convective drying at 50℃ being the temperature at which the best retention of anthocyanins was noted (as compared to drying at 60, 70, 80 and 90℃). Unfortunately, the applied drying processes significantly reduce antioxidant capacity. Therefore, using lower temperatures for finish-drying mitigates the degradation of bioactives in pomace, which suggests that such parameters are a recommendable option when it comes to drying blackcurrant pomace.

Apple pomace remaining after industrial processes was dried using the following techniques: sun drying (about 26℃), hot-air oven drying (60℃) and freeze-drying (−50℃) and then analysed in terms of selected physical properties, the presence of polyphenols, and antioxidant capacity. The highest moisture content was found in freeze-dried fractions, whereas the lowest in sun dried ones. The dietary fibre fraction obtained using the freeze-drying technique had also significantly better functional properties than the sun and oven dried fractions. Sun and oven dried fractions had the highest bulk density as a result of the shrinkage of cell wall material at elevated temperature. Water holding capacity was over twice higher after freeze-drying than after hot-air oven drying. Freeze-dried fractions displayed the highest retention of polyphenols, including flavonoids, and had better hydration properties than products dried in high temperatures. It was found out that it is those factors that affected the products' sensory attractiveness, such as colour or taste. The highest soluble-to-insoluble dietary fibre ratio was observed for hot-air oven dried fractions. The highest degradation of bioactive compounds was observed in the fraction subjected to sun drying. In conclusion, researchers pointed to relatively high potential of industrial apple pomace as a functional food ingredient after the recommended drying processes (freeze-drying and hot-air oven drying) owing to its physical and chemical properties such as the high content of dietary fibre.

Fruit and Vegetable Processing Waste Management

The processing of fruits results in generation of high amounts of waste materials such as peels, seeds, stones, pomace, rags, kernels and oilseed meals. A huge amount of waste in the form of liquid and solid is produced in the fruit and vegetable processing industry which causes pollution problem if not utilized or disposed-off properly. The waste obtained from fruit processing industry is extremely diverse due to the use of wide variety of fruits and vegetables, the broad range of processes and the multiplicity of the product. Peels are the major by-products obtained during the processing of various fruits and these were shown to be a good source of various bioactive compounds which posses various beneficial effects. But significant quantities of fruit peels (20%-30% for banana and 30%-50% for mango) are discarded as waste by the processing industries which cause a real environmental problems. Thus, new aspects concerning the use of these wastes as by-products for further exploitation on the production of food additives or supplements with high nutritional value have gained increasing interest because these are high-value products and their recovery may be economically attractive.

Nutritional Properties of Fruit and Vegetable-based Products

The nutritional roles of pectin on the human body can be divided into two stages. First, pectin is resistant to digestion and absorption within the upper gastrointestinal phase and therefore reaches the colon in a chemically unaltered state. In this stage, the pectin may modulate the digestion and absorption of nutrients by various mechanisms, including binding bile acids or calcium, altering digestive medium viscosity, changing the interface between oil and water phases, or inhibiting lipase activity. Pectin has also been proved to bind with hydrophilic polyphenols, which influences the digestion of polyphenols.

Second, the pectin is fermented in the colon by gut microbiota, resulting in the production of short chain fatty acids (SCFAs), mainly acetate, propionate and butyrate, which contribute to health benefits. During the colonic stage, pectin can regulate the composition and activity of the gut microbiota, thereby modulating the level and type of SCFAs and other bioactive metabolites produced. Pectin could increase gut microbiota, including *Bifidobacterium* spp. and butyric acid-producing bacteria, as for example *Faecalibacterium* and other members of *Ruminococcaceae* family.

(1) Nutritional properties influenced by thermal processing

The release of lipophilic and hydrophilic bioactive compounds (e.g., carotenoids and polyphenols) from raw fruit and vegetables is restricted by the cell walls, which is one of the reasons certain bioactive compounds have a low bioavailability. Generally, thermal processing increases carotenoid bioaccessibility by destroying the integrity of the cell walls and internal membranes, thereby facilitating the release of the carotenoids. Additionally, thermal treatment has been shown to solubilize, depolymerize, and demethoxylate the pectin molecules within carrot. Pectin concentration is considered to have a negative correlation with carotenoid micellization. A high pectin concentration decreased carotenoid micellization, which would be expected to reduce its bioavailability. During thermal processing, a negative correlation was observed between the texture of carrots and the bioaccessibility of β-carotene.

(2) Nutritional properties influenced by non-thermal processing

Non-thermal processing generally causes the disruption of plant cell walls, making it possible for the contact of carotenoids from chromoplast and pectin from cell wall materials. In general, studies suggest that each type of food matrix responds differently to US treatment, being influenced by processing conditions, food structure, composition, and interactions. Mechanical processing or homogenization prior to thermal processing can increase the surface area available for digestive enzymes to act, thus being an advantageous approach to increase carotenoid bioaccessibility. The beneficial effect of subsequent thermal processing following HPP, in turn, has been attributed to the fact that high temperature can weaken the physical barriers that enclose carotenoids and to the acceleration of β-elimination of pectin, which causes the softening of plant cell walls.

New Words

minimally ['mɪnəməlɪ]	*adv.*	最低限度地；最低程度地
bioactive [ˌbaɪəʊ'æktɪv]	*adj.*	生物活性的
constituent [ˌkən'stɪtuənt]	*n.*	成分；选民
	adj.	组成的；选举的
polyphenol [ˌpɒlɪ'fiːnɒl]	*n.*	多酚
biodiversity [ˌbaɪəʊdaɪ'vɜːsətɪ]	*n.*	生物多样性
pasteurization [ˌpɑːstʃəraɪ'zeɪʃn]	*n.*	加热杀菌法，巴氏杀菌法
noncovalent bond		非共价键
predominantly [prɪ'dɒmɪnəntlɪ]	*adv.*	主要地；占优势的
orifice ['ɒrɪfɪs]	*n.*	孔；穴；腔
cavitation [ˌkævɪ'teɪʃən]	*n.*	气穴现象；空穴作用；成穴
denaturation [diːˌneɪtʃə'reɪʃən]	*n.*	使变性；改变本性
ultrasound ['ʌltrəsaʊnd]	*n.*	超声；超音波
polypeptide [ˌpɒlɪ'peptaɪd]	*n.*	多肽；缩多氨酸
interaction [ˌɪntər'ækʃn]	*n.*	互动交流；相互影响；相互作用
macromolecule [ˌmækrəʊ'mɒləkjuːl]	*n.*	大分子；高分子；巨分子
pomace ['pʌmɪs]	*n.*	果渣；油渣
blackcurrant ['blækkʌrənt]	*n.*	黑醋栗；黑加仑子
bioaccessibility	*n.*	生物有效性；生物利用度
depolymerize [diː'pɒlɪməˌraɪz]	*v.*	（使）解聚
micellization [mɪselaɪ'zeɪʃən]	*n.*	胶束形成；胶束化

Notes

1) Fruit and vegetables contain a diverse range of bioactive constituents that may be beneficial to human health, including dietary fibers, vitamins, minerals, carotenoids, and

polyphenols. Eating a diet rich in fruits and vegetables may improve health and wellbeing by reducing the risk of diseases, such as cardiovascular diseases and cancer.
参考译文：水果和蔬菜含有多种可能对人类健康有益的生物活性成分，包括膳食纤维、维生素、矿物质、类胡萝卜素和多酚。食用富含水果和蔬菜的饮食可能会降低患心血管疾病和癌症等疾病的风险，从而改善健康。

2) Based on the growing demand for more minimally processed fruit and vegetable-based products, many non-thermal processing technologies have been investigated, since these innovative technologies are often less detrimental to food quality attributes than thermal processing.
参考译文：随着轻度加工果蔬产品的需求逐渐增加，开发出许多非热加工技术，因为这些创新技术往往比热加工对食品质量的损害更小。

3) The rapid change in fluid velocity and pressure when it passes through the valve generates turbulence, shear and cavitation forces that can disrupted polymers, particles or cells in the system.
参考译文：当流体通过阀门时，其速度和压力的快速变化会产生湍流、剪切和空穴力，从而破坏系统中的聚合物、颗粒或细胞。

4) The effects of US treatment are primarily ascribed to the acoustic cavitation phenomenon, which is caused by microbubble generation, growth, and implosion during the propagation of the high-intensity ultrasonic wave.
参考译文：超声处理的效果主要归因于声空化现象，这是由高强度超声波传播过程中微气泡的产生、生长和内爆引起的。

5) Food processing involves a wide variety of processing methods and parameters, but usually results in a lower quality of products in terms of reduction in the content of polyphenols (mainly anthocyanins) in fruit and vegetables.
参考译文：食品加工涉及多种加工方法和参数，但通常会降低水果和蔬菜中多酚（主要是花青素）的含量，从而导致产品质量降低。

6) One of the best-known food preservation methods is drying, which with an appropriate parameter selection may minimize losses of bioactive compounds.
参考译文：干燥是被人熟知的食品保藏方法之一，通过选择合适的参数，可以最大限度地减少生物活性化合物的损失。

7) Freeze-dried fractions displayed the highest retention of polyphenols, including flavonoids, and had better hydration properties than products dried in high temperatures.
参考译文：冻干组分对多酚（如黄酮类）的保留率最高，且水化性能优于高温干燥的产品。

8) A huge amount of waste in the form of liquid and solid is produced in the fruit and vegetable processing industry which causes pollution problem if not utilized or disposed-off properly.
参考译文：果蔬加工业产生大量的液体和固体形式的废弃物。如果不加以利用或妥善处理，就会造成污染问题。

9) First, pectin is resistant to digestion and absorption within the upper gastrointestinal phase and therefore reaches the colon in a chemically unaltered state. In this stage, the pectin may modulate the digestion and absorption of nutrients by various mechanisms, including binding bile acids or calcium, altering digestive medium viscosity, changing the interface between oil and water phases, or inhibiting lipase activity.

参考译文：首先，果胶在上消化道阶段不易被消化和吸收，因此以化学性质不变的状态到达结肠。在这个阶段，果胶可能通过多种机制调节营养物质的消化和吸收，例如结合胆汁酸或钙，改变消化介质黏度，改变油水相界面，或抑制脂肪酶活性。

10) The release of lipophilic and hydrophilic bioactive compounds (e.g., carotenoids and polyphenols) from raw fruit and vegetables is restricted by the cell walls, which is one of the reasons certain bioactive compounds have a low bioavailability.

参考译文：在水果和蔬菜中，细胞壁限制了亲脂性和亲水性生物活性化合物（如类胡萝卜素和多酚）的释放，这也是水果、蔬菜中某些生物活性化合物生物利用度低的原因之一。

Reading Material 1 Baking Food

Introduction of Bread

It is widely known that bread is an important source of carbohydrates, proteins, fibers, vitamins and minerals. Even in countries where wheat production is not enough to meet its domestic demands, as for instance Brazil and Ghana, bread is eaten almost daily. Besides this aspect, bread is also a typical food consumed by vegetarians (*n.* 素食者), vegans (*n.* 纯素主义者), and also by individuals due to their religious beliefs.

Breadmaking production and its products vary widely around the world. Basic bread recipes comprise wheat flour, water, yeast and salt. Many studies have been proposed in order to improve bread nutritional aspects, either through the use of different flours, or by increasing the level of dietary fibers, or by reducing the sodium content, between other aspects.

However, although a bread dough recipe (*n.* 食谱) can be changed in order to improve health aspects, the breadmaking process may reduce or enhance the content of antioxidant components in flours and similarly modify the bioavailability of bioactive compounds.

Moreover, the interactions in recipe ingredients (*n.* 材料) such as starch, fiber, minerals, proteins and lipids can influence the nutritional and quality characteristics of processed bread.

The quality of protein is associated to its digestion. Protein digestibility (*n.* 消化率) analyses is estimated by its intestinal absorption capacity, reflecting on the effectiveness of protein consumption on nutrition. Furthermore, protein resistance to digestion can be associated to its ability to act as allergens, and initiate the complex autoimmune celiac disease (*n.* 自身免疫性乳糜泻).

Food process can affect the predisposition (*n.* 倾向) of protein digestibility, causing protein cross-linking or denaturation (*n.* 变性), as well as modifying it as a result of interactions with other components which form the product. Researchers observed a significant increase in bread protein digestion after baking. Furthermore, they underlined that variations in protein digestibility due to baking were more noticeable than that found between two wheat varieties.

Similarly, antinutritional compounds naturally found in bread, such as oxalate (*n.* 草酸), can reduce the bioavailability of minerals like calcium and iron, and also reduce the protein digestibility. However, the breadmaking stages, like fermentation and baking, have been related to a reduction of oxalate content.

Concerning the health impact, based on bread nutrients, the major focus of attention of bread studies is related to the acrylamide (*n.* 丙烯酰胺) content. Since 1994, when the International Agency for Research on Cancer (IARC) (*n.* 国际癌症研究机构), classified acrylamide as "potentially carcinogenic (*adj.* 致癌的) to humans", acrylamide has been the subject of many studies. Researchers specifically evaluated the effect of baking conditions (time, temperature, steam usage) on acrylamide formation in breads. On the other hand, no studies were found concerning the influence of baking parameters, temperature, time and the use or not of steam at the beginning of baking, on protein digestibility, antinutrients, and other bread nutrients. This evaluation is relevant to modulate bread nutrients, while keeping color aspects, responsible for consumer acceptability. Considering the current literature available, it could be hypothesized that (a) different baking conditions can achieve final bread with different nutritional properties, i.e., (b) the higher the color variation the lower the nutritional bread properties； in addition, (c) the Maillard reaction (*n.* 美拉德反应) products, when higher color variation is observed, are responsible for the lower nutritional aspects.

Baking Technology

Baking is a form of cooking performed in an oven (*n.* 烤箱). It transforms semi-solid dough into an eatable product under the influence of heat. Bakery products are produced by straight dough method, sponge and dough method, and Chorleywood method. During baking, the dough pieces often spread, depending on the recipe used, the dough preparation, and on the applied oven conditions. Firstly, moisture removal takes place forming the structure and texture of the biscuit and in the final stage, the browning occurs. To maintain uniform color, texture, and desired final moisture content (less than 1%-2%) of the baked biscuits (*n.* 饼干), knowledge of heat and mass transfer is necessary. In biscuit production, all three modes of heat transfer play a significant role. Conduction, convection, and radiation contribute the dough expansion, moisture removal, and browning reaction respectively.

As the baking proceeds, dough changes with the temperature such as swelling of protein (at 40-50℃), protein denaturation (*n.* 蛋白质变性) (> 50℃), gluten coagulation (*n.* 面筋凝固) (> 70℃). The air or gas bubbles in the dough filled with the released evaporated water and dough get expanded as the temperature increases to some extent. Above 100℃, water held in the solution evaporates. As soon as the moisture evaporates from the dough, surface temperature rises sharply, resulting in the change in colour and when the temperature exceeds 160℃, it may decline the quality of a product. In the presence of oven-dry heat, caramelization (*n.* 变成焦糖色), dextrinization (*n.* 糊精化), and maillard reaction contribute to the development of biscuit colour. However, for different sugars, caramelization occurs at different temperatures (e.g., sucrose and glucose at 160℃； maltose at 180℃and fructose at 110℃). Dextrinisation (*n.* 糊化) of the starch occurs in the temperature range of 100-200℃.

Among the unit operations involved in biscuit manufacturing, the process ‘baking’ is an energy-consuming process. In general, large commercial bakeries use less energy per unit mass of the product, compared to small bakeries due to efficiency of scale. In the last decade, several studies on the improvement of thermal efficiency of the ovens towards the large-scale industry have been studied, but at the local or household level, it is still in progress. In the bakery industry, different types of mechanisms such as convectional (*adj.* 对流的), microwave assisting with infrared, and steam baking produce the bakery products. Although the convectional mechanism gives satisfactory baking, it results in moisture gradient and differential shrinkage due to the rapid drying of the surface. Moreover, the surface becomes an insulator to heat penetration. The microwave baking in bulk heating in which microwaves interact with material as a whole and results in more uniform moisture distribution as the heat penetrates more deeply and rapidly.

Mostly, microwave heating is used in the last stage of biscuit baking to produce products with superior properties. Researchers compared the convectional baking of leavened, canned biscuit dough with microwave baking for the textural changes and found that baking of biscuit for 50 s using microwave shows the best textural properties than conventional baking for 10 min. Moreover, biscuit baking using microwave reduces the checking to 5% compared to 61% in conventionally baked biscuits and convectional baked one becomes weak at a faster rate than microwaved biscuits. Also, a more uniform moisture distribution was observed in microwave biscuits than the convectional biscuit. Alternatively, microwaves combined with hot air or infrared radiation saves the time in which the advantage of browning and crispiness (*n.* 脆度) is due to microwave heating and halogen (*n.* 卤素) lamp heating, respectively. The combined effect of halogen lamp and microwave baking on the texture and colour of cookies was studied and compared with conventional baking (205℃, 11 min). Different levels of halogen powers (60%, 70%, and 80%), microwave powers (20% and 30%), and baking times (4.5, 5.0, 5.5, and 6.0 min) were used for baking. They found that the cookies baked at 70% power of halogen lamp and 20% microwave power levels for 5.5 min have a similar hardness and colour values as that of conventionally baked cookies.

Biscuit

Influence of Basic Ingredients on Physical Properties of Biscuit

(1) Spread ratio

Spread ratio is an important property in evaluating the quality of the biscuit. It is defined as the ratio of diameter (*n.* 直径) to height of the biscuit. In general, a biscuit with a larger spread ratio is supremely desirable. Flour with low hydration (*n.* 水合作用) properties provides a larger spread ratio to the biscuit. However, higher protein content of the flour that exhibits a greater water binding ability reduces the spread ratio of the cookie. On the contrary, the spread ratio increases for a non-wheat protein flour due to the presence of a greater amount of lipids and fibers. Similar increasing results are confirmed for amaranth (*n.* 苋菜) flour, sesame (*n.* 芝麻) peel flour, and fruit byproducts (*n.* 副产品). Moreover, the presence of the non-starch polysaccharides such as glycans, fucose, celluloses, lignin and hemicelluloses enhances the reduction in the spread ratio.

A combination of different flours has different water absorption capacities and forms aggregate. As a result, the hydrophilic (*adj.* 亲水的) sites that compete with limited water increases leading to a reduced spread ratio. The spread ratio of cookies decreases with germinated (*v.* 发芽) flour. This decrease is attributed to the enzymatic degradation of the starch and protein into smaller sugars and peptides which ultimately increase the hydrophilic nature of the cookies. Researcher incorporated the potato and corn flour in wheat flour at different levels of 2%, 4%, and 6%. They found that there is an increase in the spread ratio of cookies with an increase in proportion level due to higher dough extensibility. On the other hand, higher fat and sugar level increases the mobility of the dough on melting and improves the spread rate. However, reducing the level of sugar and fat content makes the flour accessible to the water and increases the dough consistency during baking and stops spreading.

(2) Density and thickness

Density is an index (*n.* 指标) for the sensory (*adj.* 感官的) texture of biscuits. Lower density gives greater crispiness and better texture. During baking, soft flour with less gluten content gelatinizes to a smaller extent and results in a lower density. Higher fat and sugar content hinders the accessibility of flour particles to the water and results in less development of gluten. Moreover, due to the higher amount of moisture loss, the density is less. Flour with high fiber content decreases the thickness of the biscuit. The high water absorption characteristics of fiber attract water and decrease the dough viscosity (*n.* 黏度) leading to reduced biscuit thickness. On the other hand, thickness increases in decreasing the sugar content. The increase in thickness is named as oven rise. Oven rise is caused by the water vaporization and through leavening gases.

Influence of Basic Ingredients on Textural Properties of Biscuit

From the consumer point of view, the biscuit texture should be crispy with a good mouthfeel as it determines the quality of the acceptance. In general, the texture of the biscuit is mainly related to the mechanical properties of the biscuit. In biscuit structure, gas cells of various sizes and shapes get embedded (*adj.* 嵌入…之中的) in a matrix of starch, fat, and sugar. The proportion or replacement of fat to sugar level, flour, and particle size of ingredients influence the mechanical properties of biscuits that significantly affect the texture. Biscuits prepared with wheat flour requires more force to break than biscuits made with non-wheat flour. The presence of gluten in wheat flour establishes the protein matrix and produces harder texture. Furthermore, biscuits made with germinated flour are found to be softer in texture than those containing non-germinated flour. The structural degradation of starch and protein induced by germination contributes to the formation of a weaker matrix in cookies, thus softening the texture.

Addition of sugar results in a highly cohesive (*adj.* 有凝聚力的) structure and crispy (*adj.* 松脆的) in texture. When sugar dissolves in the dough water, a high, thick viscous solution is formed and on cooling, it solidifies to become a hard, amorphous (*adj.* 无定型的), and glassy material and produces the product with crispy texture. Recently, in the study of sugar-free cookies, the addition of whey protein concentrate and emulsifier during storage, showed no significant effect on physical and textural characteristics and it could be stored up to 15 d. The type of fat substitute and its replacement level also contributes to the changes in volume and crumb firmness of the biscuit or cookie. Breaking strength, compressive strength is decreased with an increase in fat levels and improved the crispness of the biscuit. Several authors carried out extensive work on fat replacement in biscuits. For instance, cellulose (*n.* 纤维素) emulsions (*n.* 乳剂) used as a shortening replacer showed a crispy texture to the biscuit with more baking time. Moreover, an increase in the fat level increases the biscuit porosity and cell size whereas decreases the cell and cell wall anisotropy (*n.* 各向异性). Also, it results in a reduction of biscuit break strength since the fat influences the strength of the cell wall. However, the distribution of cell sizes and cell wall thickness is unaffected by the level of the fat added. Similarly, biscuit porosity, cell size, and the cell wall thickness increase with an increase in sugar level because of its effect on dough viscosity.

Another important ingredient is water content which noticeably affects the texture of the end product. Researchers investigated the effect of water content on the mechanical properties of biscuits by a three-point bending test. They found that a drop in fracture stress and modulus accompanied by an increase in deformation to fracture at a water content of about 4% to 5%. Similarly, the young's modulus of semi-sweet biscuits decreased with an increase in moisture content. Also, cracking of biscuits is usually registered due to the moisture gradients within the biscuits. During baking, biscuits have more moisture in the central region than the edge regions. However, during cooling and storage, this moisture transfers from the central region to the outer edge region. Due to this moisture migration, it leads to expansion at the outer region and contraction at the center, as a result, stresses build up within the biscuit, eventually, the cracks would develop.

The quality of the dough also significantly determines the final texture of biscuit-like hardness, fracture ability, cutting strength, crispiness, crunchiness (*n.* 脆度), brittleness, cohesiveness (*n.* 凝聚力), springiness (*n.* 弹性), chewiness. Dough that is too firm or too soft will not yield a satisfactory product. Biscuit hardness is related to the degree of partial gelatinization of starch resulting in a softer texture for less swollen starch granules. The difference between the product temperature and glass transition temperature associates the degree of crispiness.

Browning Reactions in Baking

The processes responsible for developing browning in biscuits are (a) Maillard reaction, (b) caramelization, and (c) dextrinization. These reactions develop the desired appearance or colour, flavour, texture, and taste which are important in cereal-based products. Moreover, compounds such as acrylamide and 5-(hydroxymethyl) furfural (HMF) (*n.* 羟甲基糠醛), advanced glycation products (AGEs), low-molecular-mass browning are also formed. The Maillard reaction plays a significant role in baking, frying, roasting, and other new technologies. These reactions are favoured effectively in foods when temperatures are above 50℃or a pH of 4-7. On the other hand, caramelization requires more heat treatment conditions, i.e. temperatures above 120℃ or 9 < pH < 3, and low water activity. In Maillard reaction, the condensation of amino groups and reducing sugars form intermediate chemical compounds that eventually undergo a series of reactions to form brown pigments. Several studies were carried out on the development of browning: Maillard reaction and caramelization, acrylamide formation in bakery products. During baking of cookies, the temperature and water activity are the major influencing factors for browning development. However, other parameters such as baking time, humidity (*n.* 湿度), and dough composition also affect surface colour in crackers depending on the quantities of sugar and protein content.

Notes

1) However, although a bread dough recipe can be changed in order to improve health aspects, the breadmaking process may reduce or enhance the content of antioxidant components in flours and similarly modify the bioavailability of bioactive compounds.

   参考译文：然而，尽管可以通过改变面包面团配方以改善健康状况，但面包制作过程可能会降低或提高面粉中抗氧化成分的含量，同样可能改变生物活性化合物的生物利用度。

2) Protein digestibility analyses is estimated by its intestinal absorption capacity, reflecting on the effectiveness of protein consumption on nutrition. Furthermore, protein resistance to digestion can be associated to its ability to act as allergens, and initiate the complex autoimmune celiac disease.

   参考译文：蛋白质消化率分析是通过其肠道吸收能力来评估的，反映了营养成分中蛋白质消耗的有效性。此外，蛋白质耐消化可能与其作为过敏原以及引发复杂的自身免疫性乳糜泻的能力有关。

3) Similarly, antinutritional compounds naturally found in bread, such as oxalate, can reduce the bioavailability of minerals like calcium and iron, and also reduce the protein digestibility.

参考译文：同样，面包中天然存在的抗营养物质（如草酸盐），会降低钙和铁等矿物质的生物利用率，也会降低蛋白质的消化率。

4) On the other hand, no studies were found concerning the influence of baking parameters, temperature, time and the use or not of steam at the beginning of baking, on protein digestibility, antinutrients, and other bread nutrients. This evaluation is relevant to modulate bread nutrients, while keeping color aspects, responsible for consumer acceptability.

参考译文：另一方面，并未发现关于烘焙参数、温度、时间以及烘焙开始阶段是否使用蒸汽对蛋白质消化率、抗营养物质和其他面包营养物质的影响的研究。该评估是为了调节面包营养成分，同时保持面包颜色，是对消费者的接受度负责。

5) Alternatively, microwaves combined with hot air or infrared radiation saves the time in which the advantage of browning and crispiness is due to microwave heating and halogen lamp heating, respectively.

参考译文：此外，微波与热空气或红外辐射相结合可以节省时间，因为微波加热和卤素灯加热分别具有褐变和脆化的优势。

6) Higher fat and sugar content hinders the accessibility of flour particles to the water and results in less development of gluten.

参考译文：较高的脂肪和糖含量会阻碍面粉颗粒与水的接触，并导致面筋的生成量减少。

7) The proportion or replacement of fat to sugar level, flour, and particle size of ingredients influence the mechanical properties of biscuits that significantly affect the texture.

参考译文：脂肪与糖的比例或替代量、面粉和原料颗粒大小会影响饼干的机械性能，从而显著影响饼干的质地。

8) Recently, in the study of sugar-free cookies, the addition of whey protein concentrate and emulsifier during storage, showed no significant effect on physical and textural characteristics and it could be stored up to 15 d.

参考译文：近期，在无糖饼干的研究中，添加乳清蛋白浓缩物和乳化剂对饼干的物理和质地特性没有显著影响，贮存期长达 15 天。

9) Biscuit hardness is related to the degree of partial gelatinization of starch resulting in a softer texture for less swollen starch granules.

参考译文：饼干硬度与淀粉的部分糊化程度有关，从而使较低膨胀程度的淀粉颗粒具有较软的质地。

Exercise

1. Translate the following sentences into Chinese.

1) Many studies have been proposed in order to improve bread nutritional aspects, either through the use of different flours, or by increasing the level of dietary fibers, or by reducing the sodium content, between other aspects.

2) The quality of the dough also significantly determines the final texture of biscuit-like hardness, fracture ability, cutting strength, crispiness, crunchiness, brittleness, cohesiveness, springiness, chewiness.

3) The processes responsible for developing browning in biscuits are (a) Maillard reaction, (b) caramelization, and (c) dextrinization. These reactions develop the desired appearance or colour, flavour, texture, and taste which are important in cereal-based products.

2. Translate the following sentences into English.

1）众所周知，面包是碳水化合物、蛋白质、膳食纤维、维生素和矿物质的重要来源。

2）首先，去除水分形成饼干的结构和质地，在饼干制作的最后阶段，发生褐变反应。

3）另一方面，随着糖含量的降低，饼干厚度增加。

4）类似地，饼干孔隙率、孔大小和孔壁厚度随着糖含量的增加而增加，因为它对面团黏度有影响。

Reading Material 2 Baby Food

Breast feeding of young babies is undoubtedly preferable to ensure the development of a healthy child. However, breast feeding is not always possible, and then the baby food should be offered a surrogate. Unmodified cow's milk is definitely unsuitable, a reason why specific infant formulas have been developed. These are for the most part based on fractions of cow's milk.

Composition of Human Milk

The composition of human milk strongly differs from that of cow's milk. It should be noticed that composition of human milk greatly varies, especially among individual mothers. The milk also changes strongly throughout the lactation period; the 'mature' milk (after two weeks) is considerably different from colostrum (milk of the first few days after birth). Table 5.2 is incomplete: the components given are important for nutrition or differ substantially between human and bovine milk. The table also shows minimum amounts of nutrients recommended for young babies.

Table 5.2 Composition[a] of human milk and cow's milk and minimum requirements for infant formulas[b]

Component	Unit	Human olostrum[c]	Human milk[c]	Cow's milk[c]	Requirement per 300 kJ
Energy	kJ	240	300	290	
Fat	g	2.5	4.2	4.0	3
Linoleic acid	mg		400	70	200
α-Linoleic acid	mg		40	15	35
Cholesterol	mg	25	20	13[d]	

（continued）

Component	Unit	Human olostrum[c]	Human milk[c]	Cow's milk[c]	Requirement per 300 kJ
Lactose	g	5.0	6.3	4.6	5
Other saccharides	g	1.8	1.3	0.1	
Protein	g	1.6	0.8	3.3	13
NPN compounds	g	0.5	0.5	0.1	
Calcium	mg	30	35	115	35
Magnesium	mg	3.5	3	11	4
Zinc	mg	1	0.3	0.4	0.35
Iron	μg	75	80	20	140
Copper	μg	60	40	2	30
Phosphorus	mg	14	14	95	18
Iodine	μg		7	5	3.5
Vitamin A	RE[e]	200	80	45	140
Vitamin D	μg		0.01	0.06	0.7
Vitamin E	mg	1	0.4	0.1	0.5
Thiamin B_1	μg	2	17	45	30
Riboflavin B_2	μg	30	30	180	40
Niacin	mg	0.06	0.2	0.5	0.2-0.6
Vitamin B_6	μg		6	65	25
Folic acid	μg		5	5	3
Vitamin B_{12}	μg		0.01	0.4	0.1
Ascorbic acid	mg	4	4	2	6

Note: a. Not complete, approximate.; b. For babies below 6 months old.; c. Amounts per 100 g milk.; d. Cow's skim milk: 2 mg; e. Retinol equivalent = μg retinol + μg carotene/6.

Lipids: The total lipid content is as in cow's milk, but the fatty acid residues in the triglycerides show a quite different pattern. Short-chain acids (fewer than 12 C-atoms) are hardly present, and the fat contains great amounts of polyunsaturated fatty acids: 18-22 C-atoms, 2-5 double bonds. The content of 'essential' fatty acids is much higher than in cow's milk. Most palmitic acid (*n.* 棕榈酸) is esterified in 2-position, which means that upon lipolysis in the gut a glycerol monopalmitate results, which is readily absorbed by babies, unlike the stearic and palmitic acids resulting from the digestion of bovine milk fat. Hence, the fat used for infant formulas generally consists for the most part of suitable vegetable oils. The latter implies that infant formulas are very poor in cholesterol (*n.* 胆固醇) (about 1 mg/100 mL), as compared to human milk (see Table 5.1).

Carbohydrates: Human milk contains, besides a relatively high amount of lactose, a substantial of oligosaccharides (*n.* 低聚糖). These have between 3 and 14 saccharide units, and most of these have a lactose residue and some *N*-acetyl groups. Their function is not yet fully clear, but it is assumed that some oligosaccharides promote growth of certain bifidobacteria in the large intestine, as these compounds cannot be hydrolyzed by the native enzymes in the gut. Oligosaccharides from various sources are added to some infant formulas; in other cases, lactulose is added, which also stimulates the growth of bifidobacteria (*n.* 双歧杆菌).

Protein: See Table 5.3 for protein composition. When feeding a baby with cow's milk, its kidneys have difficulty in processing the large amounts of degradation products of the protein metabolism, especially in combination with the large amount of minerals going along with the casein. Moreover, the casein of cow's milk gives a firm clot in the stomach, and it takes a long time before the small intestine. These problems do not arise with human milk, due to its far lower protein content and small proportion of casein (about 30% of the protein as compared to 80% in cow's milk). When making infant formulas from cow's milk, the protein composition thus needs considerable adjustment.

Table 5.3 Protein in human and cow's milk

Protein	Human colostrum	Human milk, nature	Cow's milk
Casein	5[a]	2.5[a]	26
α-Lactalbumin	3	2	1.2
β-Lactoglobulin	0.0	0.0	3.2
Serum albumin	0.4	0.3	0.4
Immunoglobulins	2.5	0.8	0.7
Lactoferrin	3.5	1.5	<0.1
Lysozyme	0.5	0.5	10^{-4}

Note: Approximate average in grams per kg; incomplete.

a. Predominantly β- and κ-casein.

The composition of the serum protein also differs. Striking is the absence of β-lactoglobulin from human milk, and the presence of a large proportion of antimicrobial proteins, notably immunoglobulin A, lysozyme, and lactoferrin. The amino acid composition of human and bovine milk proteins is not significantly different. Finally, human milk contains a long amount of non-protein nitrogen. The functions of these compounds are largely unclear; much of the material is indigestible.

Mineral: The content of minerals (inorganic) in human milk (about 0.2%) is much lower than in cow's milk (0.6%-0.7%). This is in accordance with the low protein and the high lactose content. The contents of some trace elements, notably iron and copper, are relatively high.

The lower content of protein and calcium phosphate in human as compared to bovine milk is clearly related to the relative growth rate of a baby being much slower than that of a calf.

Vitamins: For the most part, the differences between human and cow's milk are fairly small, but the content of some vitamins is much higher in cow's milk. This seems to pose no problems.

Formula Composition and Manufacture

The composition of the formula should comply with recommended quantities (the recommendations do not greatly differ between counties). Generally, skim milk and sweet whey are used, ratio, e.g., 1 to 5. The whey should not contain more than traces of lactic acid and no added nitrate. Part of the whey should be desalted. Carbohydrate content can be boosted by addition of lactose or a UF permeate. Sometimes, oligosaccharides are added (or lactulose). The fat is generally a mixture of vegetable oils, containing sufficient fatty acids and oil-soluble vitamins. Vitamin C is usually added, and if necessary, vitamin A, D and E. Fortification with Fe and Cu are commonly practiced.

This all concerns a formula for healthy babies of up to 6 months old. For babies over 6 months old, the composition of the mix is different, generally involving a much greater proportion of (skim) milk. Other products are needed for children born preterm, or for children suffering from allergy or some metabolic disease. The manufacture generally involves wet mixing of the ingredients and pre-emulsification, followed by pasteurization and homogenization. Occasionally, an emulsifier is added, but this is not necessary. After pasteurization either a liquid or a dried product can be made. For the former, the milk is UHT-sterilized, followed by aseptic packing in cartons. Concentrated products that are sterilized in bottles or cans are also produced. When making powdered formulas, the milk is concentrated by evaporation, followed by spray drying; the pasteurization should be sufficiently intense to kill virtually all pathogens.

Notes

1) Most palmitic acid is esterified in 2-position, which means that upon lipolysis in the gut a glycerol monopalmitate results, which is readily absorbed by babies, unlike the stearic and palmitic acids resulting from the digestion of bovine milk fat. Hence, the fat used for infant formulas generally consists for the most part of suitable vegetable oils.

   参考译文：大多数棕榈酸在 2 位上进行酯化，这意味着脂肪在肠道中分解为很容易被婴儿吸收的甘油单棕榈酸酯，这与牛奶脂肪消化得到硬脂酸和棕榈酸不同。因此，用于婴儿食品的脂肪通常包括合适的植物油的大部分。

2) Their function is not yet fully clear, but it is assumed that some oligosaccharides promote growth of certain bifidobacteria in the large intestine, as these compounds cannot be hydrolyzed by the native enzymes in the gut.

   参考译文：它们的功能还不是完全清楚，但是由于这些化合物不能被肠道内的天然酶水解，故可以设想一些寡糖促进了特定的双歧杆菌在大肠内的生长。

Exercise

True or false.

1) The whey to casein ratio of mature human milk is 70:30 and that of cow milk are 20:80. ( )

2) For normal babies, infant formula could meet the nutritional needs of 0-12 months old babies. ( )

3) Compared with breast milk, cow's milk has a low content of lactose and whey protein, and also a high level of casein. ( )

Reading Material 3 Quick-Freezing of Foods

Introduction

Freezing is one of the more common processes for the preservation of foods. A further principle underlying the modern process of quick-freezing is that the rapid withdrawal of heat leads to the conversion of the watery fraction of the food into very small crystals which do much less harm to the structure of the commodity frozen than the larger crystals produced when the rate of freezing is slow.

The Freezing Points of Foods

The freezing point of a liquid is that temperature at which the liquid is in equilibrium with the solid. A solution with a vapor pressure lower than that of a pure solvent will not be in equilibrium with the solid solvent as its normal freezing point. The system must be cooled to that temperature at which the solution and the solid solvent have the same vapor pressure. The freezing point of a solution is lower than that of a pure solvent. The freezing point of food is lower than that of pure water.

Percent Water Frozen vs. Temperature of Food and its Quality

Careful evaluation of the freezing curve for a food, beef for example, under controlled conditions, demonstrates first that super-cooling occurs, and that this is characteristic for products. With a thin section of food tissue, it can be shown that following super-cooling the temperature of the cooled section rises to the actual freezing point when the change in phase occurs. This change in phase continues, providing a temperature differential is maintained, until the free water becomes ice.

The water in foods exists in two (or more) states. The generally used terms are 'bound' water and 'free' water. Flavor changes, color changes, nutrient losses, and texture losses occur relatively rapidly above 15°F (as compared to 0°F or lower). The lower the temperature, the slower the rate of loss of ascorbic acid. Further, some products deteriorate more rapidly under fluctuating temperatures.

Volume Changes during Freezing

In a freezing environment an unprotected bottle of milk freezes, and, if it does, it usually pushes the cap from the bottle. The milk expands on freezing; the volume of milk increases. The cap sits on top of a protruding cylinder of frozen milk. If the milk bottle has a solidly held cap of metal, one that would resist being pushed from the bottle, then it would burst. The liquid is not compressible, and neither is the ice. As the volume increase within the bottle during freezing, either the bottle bursts or the cap is forced free.

When freezing food in a rigid container, opportunity for such expansion must be considered. However, not all food products expand during freezing. Strawberry jam does not increase in volume when frozen.

Refrigeration Requirements in Freezing Foods

Hot and cold are relative concepts, requiring a reference point. The term ice carries information, and in its common usage, refers to the solid state of water molecules. Water freezes below 32°F. To keep water molecules in the solid state, it is necessary to have them in an environment which permits the solid state to exist.

First, a condition must be established such that heat can be removed from liquid water, and a change in state permitted to occur. Next, the frozen material must be protected (insulated) to prevent it from acquiring heat, raising the temperature, and thawing the ice formed. There are two distinct problems areas; the problem of bringing a sample to a frozen condition, and the problem of maintaining the frozen material in a suitable solid state. In order to establish the refrigeration requirements to achieve this desired state, it is necessary to consider both aspects of the problem.

Influence of Freezing on Microorganisms

Microorganisms may be classified by their optimum temperatures for growth. Most microorganisms do not grow at temperatures below 32°F. However, it is known that there are some yeasts that can grow at temperatures as low as 15°F in non-frozen substrates. In general, yeasts and molds are capable of growing at temperatures much lower than those supporting the growth of bacteria. Slow freezing is damaging to a microbial population. The most susceptible forms of microorganisms are the vegetative cells. Spores are not usually injured by freezing.

The temperatures at which frozen food is thawed also have a marked influence on the growth of microorganisms.

Influence of Freezing on Food Quality

Enzyme activity is temperature dependent. The activity of an enzyme or a system of enzymes can be destroyed at temperatures near 200°F. Enzymes retain some activity at temperatures as low as −100°F, although reaction rates are extremely low at that temperature. Animal enzyme systems tend to have optimum reaction rates at temperatures near 98°F. Plant enzyme systems tend to have temperature optimal at slightly lower temperatures.

Freezing stops microbiological activity. Enzyme activity is only retarded by freezing temperatures. Enzyme control is easiest obtained by destroying them with a short heat treatment (blanching) prior to freezing and storage.

Oxidative deteriorations of fats and oils are not uncommon in frozen foods rich in these nutrients. Fatty fish deterioration is notable. Fats in frozen fish tissue tend to become rancid quicker than the fats in frozen animal tissues. Plant tissues are least susceptible.

Emulsions of oil in water or water in oil may become destabilized by freezing. This may be serious in prepared precooked frozen foods and foods products. Because fat and oil deteriorations are, in general, temperature dependent, freezing preservation offers maximum potential in preserving many fatty foods.

During the processing steps, nutrient losses may occur. Losses of vitamins occur throughout processing operations, for example, during blanching and washing, trimming and grinding. Exposure of tissues to the atmosphere results in vitamin losses due to oxidation. In general, vitamin C losses will occur when tissues are ruptured and exposed to the air.

The storage of foods in a frozen state without package protection leads to oxidation and destruction of many nutrients, including vitamins.

Thawing Damage to Frozen Foods

Refreezing a thawed mass of food may result in important quality changes, and these are measurable to some extent. The changes in the freezing point of thawed frozen foods may well be a measure of the changes which occur in freezing plant and animal tissues.

Animal tissues appear less prone to freezing and thawing damage than are fruit and vegetable tissues, and fruits are particularly sensitive.

If frozen concentrated orange juice is processed quickly and carefully, but allowed to thaw and freeze several times during distribution, the thawed juice when consumed will little resemble the juice of fresh oranges in odor, taste or appearance. Repeated freezing and thawing are detrimental to frozen fruits, vegetables, prepared foods, ice cream, fish, and poultry.

Notes

1) A solution with a vapor pressure lower than that of a pure solvent will not be in equilibrium with the solid solvent at its normal freezing point. The system must be cooled to that temperature at which the solution and the solid solvent have the same vapor pressure.

参考译文：蒸汽压低于纯溶剂的某溶液在正常凝固点时与固体溶剂不能达到平衡。该系统必须冷却到溶液和固体溶剂具有同样蒸汽压的温度下（方能达到平衡）。

2) With a thin section of food tissue it can be shown that following super-cooling the temperature of the cooled section rises to the actual freezing point when the change in phase occurs. This change in phase continues, providing a temperature differential is maintained until the free water becomes ice.

参考译文：用一片薄的食品组织可以看到，在过冷现象后，当相变发生时，冷冻组织的温度上升到实际凝固点。如果温度差得以维持，则这种相变持续发生，直到所有的自由水变成冰。

Exercise

1. Translate the following sentences into English.

1) 液体的凝固点是液体与固体达到平衡时的温度。

2) 一般来说，酵母和霉菌能够在比细菌生长温度低得多的温度下生长。

2. Translate the following sentences into Chinese.

The water in foods exists in two (or more) states. The generally used terms are 'bound' water and 'free' water. Flavor changes, color changes, nutrient losses, and texture losses occur relatively rapidly above 15°F (as compared to 0°F or lower). The lower the temperature, the slower the rate of loss of ascorbic acid.

3. True or false.

1) Most microorganisms do not grow at temperatures below 32°F. (   )

2）Enzymes retain some activity at temperatures as low as −100°F, although reaction rates are extremely low at that temperature. (   )

Chapter 6

Food Quality and Safety

Lesson 1 Food Quality and Control

Introduction

With the deepening of China's reform and opening up, the development of modern science and technology, the improvement of people's living standards, people's requirements for food quality are also improving, the production and processing channels of food are also increasing, food quality and safety have become an important aspect to ensure people's livelihood. With the factors of enterprise leading and market needing, the food production and development of our country has entered a new stage, and gradually developed into the pillar industry of national economy development. In the process of food production and development, in addition to the urgent need to improve the efficiency of food production, but also faced with increasingly severe food safety adjustment. In developed countries, the EU, Japan, the US and other developed countries have witnessed various food production safety incidents, such as foot-and-mouth disease, mad cow disease and other malignant food safety incidents, these food safety incidents have brought very serious public health safety. These food safety problems also have a very bad impact on the food market, which makes the image of the whole industry seriously damaged, reduces the competitiveness of the industry, and threatens the benefits and development of the food industry.

Food can transmit pathogens which can result in the illness or death of the person or other animals. The main types of pathogens are bacteria, viruses, mold, and fungus. Food can also serve as a growth and reproductive medium for pathogens. In developed countries there are intricate standards for food preparation, whereas in lesser developed countries there are fewer standards and less enforcement of those standards. Another main issue is simply the availability of adequate safe water, which is usually a critical item in the spreading of diseases. In theory, food poisoning is 100% preventable. However, this cannot be achieved due to the number of persons involved in the supply chain, as well as the fact that pathogens can be introduced into foods no matter how many precautions are taken.

Food safety (or food hygiene) is used as a scientific method/discipline describing handling, preparation, and storage of food in ways that prevent food-borne illness. The occurrence of two or more cases of a similar illnesses resulting from the ingestion of a common food is known as a food-borne disease outbreak. This includes a number of routines that should be followed to avoid potential health hazards. In this way, food safety often overlaps with food defense to prevent harm to consumers. The tracks within this line of thought are safety between industry and the market and then between the market and the consumer. In considering industry to market practices, food safety considerations include the origins of food including the practices relating to food labeling, food hygiene, food additives and pesticide residues, as well as policies on biotechnology and food, guidelines for the management of governmental import and export inspection and certification systems for foods. In considering market to consumer practices, the usual thought is that food ought to be safe in the market and the concern is safe delivery and preparation of the food for the consumer.

Introduction of GMP

GMP are the practices required in order to conform to the guidelines recommended by agencies that control the authorization and licensing of the manufacture and sale of food and beverages, cosmetics, pharmaceutical products, dietary supplements, and medical devices. These guidelines provide minimum requirements that a manufacturer must meet to assure that their products are consistently high in quality, from batch to batch, for their intended use. The rules that govern each industry may differ significantly; however, the main purpose of GMP is always to prevent harm from occurring to the end user. Additional tenets include ensuring the end product is free from contamination, that it is consistent in its manufacture, that its manufacture has been well documented, that personnel are well trained, and that the product has been checked for quality more than just at the end phase. GMP is typically ensured through the effective use of a quality management system (QMS).

GMP guidelines provide guidance for manufacturing, testing, and quality assurance in order to ensure that a manufactured product is safe for human consumption or use. Many countries have legislated that manufacturers follow GMP procedures and create their own GMP guidelines that correspond with their legislation.

Basic Prints of GMP List as Follows:

- Manufacturing facilities must maintain a clean and hygienic manufacturing area.
- Manufacturing facilities must maintain controlled environmental conditions in order to prevent cross-contamination from adulterants and allergens that may render the product unsafe for human consumption or use.
- Manufacturing processes must be clearly defined and controlled. All critical processes are validated to ensure consistency and compliance with specifications.
- Manufacturing processes must be controlled, and any changes to the process must be

evaluated. Changes that affect the quality of the drug are validated as necessary.

- Instructions and procedures must be written in clear and language using good documentation practices.
- Operators must be trained to carry out and document procedures.
- Records must be made, manually or electronically, during manufacture that demonstrate that all the steps required by the defined procedures and instructions were in fact taken and that the quantity and quality of the food or drug was as expected. Deviations must be investigated and documented.
- Records of manufacture (including distribution) that enable the complete history of a batch to be traced must be retained in a comprehensible and accessible form.
- Any distribution of products must minimize any risk to their quality.
- A system must be in place for recalling any batch from sale or supply.
- Complaints about marketed products must be examined, the causes of quality defects must be investigated, and appropriate measures must be taken with respect to the defective products and to prevent recurrence.

Introduction of SSOPs

Sanitation Standard Operating Procedures (SSOPs) is the common name given to the sanitation procedures in food production plants which are required by the Food Safety and Inspection Service of the USDA and regulated by 9 CFR part 416 in conjunction with 21 CFR part 178.1010. It is considered one of the prerequisite programs of HACCP.

SSOPs are generally documented steps that must be followed to ensure adequate cleaning of product contact and non-product surfaces. These cleaning procedures must be detailed enough to make certain that adulteration of product will not occur. All HACCP plans require SSOPs to be documented and reviewed periodically to incorporate changes to the physical plant. This reviewing procedure can take on many forms, from annual formal reviews to random reviews, but any review should be done by "responsible educated management". As these procedures can make their way into the public record if there are serious failures, they might be looked at as public documents because they are required by the government. SSOPs, in conjunction with the Master Sanitation Schedule and Pre-Operational Inspection Program, form the entire sanitation operational guidelines for food-related processing and one of the primary backbones of all food industry HACCP plans.

SSOPs can be very simple to extremely intricate depending on the focus. Food industry equipment should be constructed of sanitary design; however, some automated processing equipment by necessity is difficult to clean. An individual SSOP should include:

- The equipment or affected area to be cleaned, identified by common name.
- The tools necessary to prepare the equipment or area to be cleaned.
- How to disassemble the area or equipment.

- The method of cleaning and sanitizing.
- SSOPs can be standalone documents, but they should also serve as work instructions as this will help ensure they are accurate.

Packaging

Packaging is the last step in food processing, however it is one of the most important steps as it has great influence on the storage life of the food product. The main purpose of packaging is to protect the food material from the external environmental hazards. It not only maintains the quality of the food product but also provide information about the ingredients and aids in distribution and marketing of the product to the final consumer. Of all these, preserving the quality of the packaged food is the main aim of packaging as quality maintenance of the food product is the most critical issue during the whole supply chain. In the last few years, new techniques such as modified atmosphere packaging (MAP), edible coating, antimicrobial packaging and antioxidant packaging have been developed. These techniques play a significant role in extending the storage life and maintaining the quality of a variety of fresh and processed food products to meet the need of the consumers.

Since the successful development and commercialization of MAP technology several decades ago, MAP is largely used for packaging different food products, which mainly include fruit and vegetables, muscle foods, dairy foods, bakery products, ready meals and dried foods.

On the other hand, intelligent packaging is a new emerging technique in the food packaging circle, protecting the food material as well as informing the consumer about the environmental condition of the packaged food. It should be noted that the concept of intelligent packaging is closely related to that of active packaging. Active packaging usually means that the package has active functions beyond the inert passive containment and protection of the food product, while intelligent packaging emphases the ability to sense or measure an attribute of the packaged food product, the atmosphere inside the package, or the environment of shipping. The information received from intelligent packaging can be communicated to users or can trigger the functions of active packaging.

New Words

foot-and-mouth disease		口蹄疫
malignant [mə'lɪgnənt]	*adj.*	恶性的；恶意的；恶毒的
intricate ['ɪntrɪkət]	*adj.*	错综复杂的
poison ['pɔɪzən]	*n.*	中毒；毒害
food-borne disease	*n.*	食源性疾病
precaution [prɪ'kɔːʃn]	*n.*	预防措施；预防；防备

food additive	*n.*	食品添加剂
pesticide residue	*n.*	农药残留
good manufacturing practice (GMP)		良好生产规范
pharmaceutical [ˌfɑːmə'suːtɪkl]	*adj.*	制药的；配药的
	n.	药品；成药
batch [bætʃ]	*n.*	一批
	vt.	一批生产量；分批处理
legislate ['ledʒɪsleɪt]	*v.*	制定法律；立法
cross-contamination	*n.*	交叉污染
allergen ['ælədʒən]	*n.*	过敏原
sanitation standard operating procedure (SSOP)		卫生标准操作规范
document ['dɒkjumənt]	*n.*	文件；文献；证件；单据；文档
	vt.	记录
conjunction [kən'dʒʌŋkʃn]	*n.*	结合；同时发生；关联
significant [sɪg'nɪfɪkənt]	*adj.*	重要的；显著的；有重大意义的
modified atmosphere packaging（MAP）		气调包装
intelligent packaging		智能包装

Notes

1) With the deepening of China's reform and opening up, the development of modern science and technology and the improvement of people's living standards, people's requirements for food quality are also improving, and the production and processing channels of food are also increasing, and food quality and safety have become an important aspect to ensure people's livelihood.

   参考译文：随着我国改革开放的深入，现代科学技术的发展和人民生活水平的提高，人们对食品质量的要求也在提高，食品的生产加工渠道也在增加，食品质量安全已成为保障民生的重要方面。

2) In considering industry to market practices, food safety considerations include the origins of food including the practices relating to food labeling, food hygiene, food additives and pesticide residues, as well as policies on biotechnology and food, guidelines for the management of governmental import and export inspection and certification systems for foods.

   参考译文：从工业生产到市场实践过程中，食品安全需要考虑的因素包括食品来源（包括与食品标签有关的规范）、食品卫生、食品添加剂和农药残留，以及生物技术和食品的相关政策、政府对食品进出口检验和认证系统的管理指南。

3) GMP are the practices required in order to conform to the guidelines recommended by agencies that control the authorization and licensing of the manufacture and sale of food and beverages, cosmetics, pharmaceutical products, dietary supplements, and medical devices.

参考译文：良好生产规范（GMP）是法规要求，目的是符合管理机构的推荐指南。这些机构能够管理食品和饮料、化妆品、医药产品、膳食补充剂和医疗器械的生产和销售的授权和许可。

4) Additional tenets include ensuring the end product is free from contamination, that it is consistent in its manufacture, that its manufacture has been well documented, that personnel are well trained, and that the product has been checked for quality more than just at the end phase.

参考译文：其他原则包括确保最终产品不受污染，与生产过程保持一致，产品制造过程有完整的文档记录，生产人员受过良好的培训，以及产品的质量检查不仅只在最后阶段进行。

5) GMP guidelines provide guidance for manufacturing, testing, and quality assurance in order to ensure that a manufactured product is safe for human consumption or use.

参考译文：良好生产规范指南为产品的制造、测试和质量保证提供指导，以确保生产的产品可供消费者安全使用。

6) Sanitation Standard Operating Procedures (SSOPs) is the common name given to the sanitation procedures in food production plants which are required by the Food Safety and Inspection Service of the USDA and regulated by 9 CFR part 416 in conjunction with 21 CFR part 178.1010. It is considered one of the prerequisite programs of HACCP.

参考译文：卫生标准操作规范是美国农业部食品安全和检验局要求的食品生产厂卫生程序的通用名称，并受第 9 章 416 款和第 21 章 178.1010 部分的规定。它被认为是 HACCP 的前提方案之一。

7) Packaging is the last step in food processing, however it is one of the most important steps as it has great influence on the storage life of the food product. The main purpose of packaging is to protect the food material from the external environmental hazards. It not only maintains the quality of the food product but also provide information about the ingredients and aids in distribution and marketing of the product to the final consumer.

参考译文：包装是食品加工的最后一步，也是最重要的步骤之一，因为它对食品的贮藏期有很大影响。包装的主要目的是保护食品材料免受外部环境危害。它不仅保持食品的质量，而且在产品向最终消费者的分销和营销过程中，提供有关食品成分和加工助剂的信息。

8) On the other hand, intelligent packaging is a new emerging technique in the food packaging circle, protecting the food material as well as informing the consumer about the environmental condition of the packaged food.

参考译文：另一方面，智能包装是食品包装的一项新兴技术，既保护了食品原料，又让消费者了解了包装食品的环境条件。

Lesson 2 Food Quality Testing

Introduction

In recent years, as a result of improvements in living standards, it has been required that food is not only fresh and delicious, but also that it meets safety and quality standards. In order to improve the taste, quality, and storage life of food, natural and artificial additives are often added. Food safety directly affects health, and food quality detection is of great significance. The WHO has clear regulations regarding the amounts of various added ingredients that can be present in food. Levels of food additives in line with national or international standards are harmless to the human body, but excessive levels of food additives or illegal additives can be harmful to the human body. For example, they may cause adverse reactions, such as allergies and poisoning, even directly leading to death in serious cases. In addition, the improper storage of food or other circumstances may lead to mildew, producing toxic and harmful bacteria and toxins. This will also cause food safety problems. At present, the detection of target substances in food is mainly carried out via high performance liquid chromatography (HPLC), gas chromatography (GC), gas chromatography-mass spectrometry (GC-MC), and other methods. These methods have advantages such as high sensitivity and low detection limits, but they often need complex instruments and complex sample processing processes, and they are high-cost.

Metabolomics

Metabolomics is one of the most important part of omics techniques that studies small molecules or metabolites within food, organisms, plants or humans. Together with genomics, transcriptomics and proteomics, metabolomics is involved in the study of food omics approaches among others. It has become an important tool in many research areas in the last few years, as food science, that involved food quality, food safety and food traceability. Therefore, metabolomics approach is a suitable solution to ensure food safety, which has become a major issue worldwide, and contaminants (i.e., organic pollutants) or toxins should be monitored in order to satisfy the consumer demand. Food safety involves several issues, including the legislation regarding the presence of selected compounds in foods that may be present below certain limits [maximum residue limits (MRLs)], the detection of microbial spoilage, environmental contaminants as well as banned external compounds, or natural toxins. In the field of food safety, liquid chromatography (LC) and GC coupled to mass spectrometry (MS) are the most important techniques used. They are powerful tools for the detection, assignment, quantification and elucidation of compounds/structures in complex matrices. GC-MS is used for the analysis of volatile metabolites whereas LC-MS is suitable for the analysis of semi-polar and polar compounds. When MS is used as detection system, low resolution mass spectrometry (LRMS) and high resolution mass spectrometry (HRMS) can be used. LRMS provides information about the nominal mass of the analyte and HRMS gives the *m/z* with four to six decimal digits. Although both LRMS and HRMS can be used for metabolomics studies, information provided by LRMS was limited, because only analytes included in a list are usually monitored. In metabolomics studies, MS coupled to LC and GC used two analysis strategies: targeted or untargeted. In the first case a list of known compounds are firstly selected, for example pesticides and their known metabolites in multiresidue methods. In contrast, untargeted analysis focuses on the detection of as many groups of metabolites as possible to obtain patterns or fingerprints without identifying nor quantifying specific compounds.

Untargeted analyses are most interesting and important in the field of metabolomics, and two strategies can be applied: fingerprinting and profiling. Fingerprinting is a fast, convenient and effective tool to classify samples based on metabolite patterns and it identifies the most important regions of the spectrum for further analysis. This approach utilizes chemometric tools as principal component analysis (PCA) or partial least square discriminant analysis (PLS-DA) to classify the components obtained in the samples in different regions to obtain a typical fingerprint of the sample. In contrast, profiling is based on the identification of the most discriminant metabolites detected in samples using databases or software tools. In addition, profiling can also include targeted analysis, monitoring a very limited number of metabolites, such as analytes or products of biochemical reactions. Other technique recently used in the field of metabolomics is nuclear magnetic resonance (NMR) spectroscopy. When it is associated with chemometric techniques, this has the potential to elucidate the interactions between targeted analytes and the environment. NMR has several advantages as straightforward sample preparation, high sample throughput, stable chemical shifts, quantification without standards, and reliable identification of isolate metabolites. However, it has several shortcomings compared with chromatographic techniques coupled to MS. For example, NMR has a limited sensitivity (compounds with concentration lower than mg/kg are not detected) and the identification in complex mixtures is difficult.

Nanozymes

Based on the unique advantages of nanomaterials and their similar catalytic activities to enzymes, nanozymes have been gradually applied in the field of food safety detection. Via catalyzing a substrate to generate a colored product, nanozyme-based systems can be used for the colorimetric determination of target substances in food. For food safety detection, compared with natural enzymes, nanozymes are relatively simple in terms of their preparation and purification, not requiring complex equipment or instruments. In addition, they can be produced in large quantities at relatively low cost. In terms of recycling efficiency and stability, nanozymes also have more advantages than natural enzymes. Moreover, some food analysis involves the modification of antibodies or aptamers, which may cause a decrease in the catalytic activities of natural enzymes. Until now, nanozymes for sensing have mainly involved biological detection, such as the detection of target substances in serum or solution.

Foodomics

The food quality assurance before selling is a needed requirement intended for protecting consumer interests. In the same way, it is also indispensable to promote continuous improvement of sensory and nutritional properties. In this regard, food research has recently contributed with studies focused on the use of 'foodomics'. This part focuses on the use of this technology, represented by transcriptomics, proteomics, and metabolomics, for the control and quality improvement of dairy products.

The especially nutrient-rich matrix of dairy products is an excellent culture medium for the growth of hazardous microorganisms, such as *Brucella abortus*, *Brucella melitensis*, *Campylobacter jejuni*, *Escherichia coli*, *Listeria monocytogenes*, *Mycobacterium bovis*, *Mycobacterium tuberculosis*, *Salmonella*, *Staphyloccocus aureus*, and *Yersinia enterocolitica*, which can seriously deteriorate the health of certain groups, such as children, pregnant women, and elderly. Transcriptomics could be very useful in predicting the activity of these pathogenic microorganisms in dairy products. It is possible to understand microbial behavior in different exogenous conditions through the analysis of RNA molecules, since the food matrix plays an important role in the expression of genes involved in activities, such as growth, survival, or level of virulence towards the host. Researchers used the transcriptomic technology to assess the extent to which a cheese matrix could modulate the expression of virulence genes in *Staphylococcus aureus*. After studying the behavior of the bacterium without and in the company of the endogenous bacterium *Lactococcus lactis*, a dominant species in the cheese ecosystem, the researchers observed that, depending on the presence or absence of the latter bacterium, there is an upward or downward production of certain enterotoxins. Thus, for example, the expression of the sea toxin was slightly favored by the presence of *Lactococcus lactis*, on the contrary that the sec4 toxin, which was down-regulated. On the other hand, the low pH caused by the presence of the endogenous bacterium was one of the factors behind the down regulation of the accessory gene regulator system, a key regulator of bacterial virulence. Therefore, in view of these results, transcriptomic studies might open new ways for the development of novel prevention strategies against the main foodborne pathogens. Instead of trying to predict the possible virulence of certain species in food, its early detection may be more interesting to prevent any possible episode or event of an infectious nature.

The study of the proteome of microorganisms can be an efficient strategy for the rapid detection of pathogens in food samples. The technique matrix-assisted laser desorption/ionization and subsequent analysis of the ions produced by time-of-flight (MALDI-TOF) seems to have fulfilled the expectations of the researches, providing satisfactory results. The target of applying this technique is the obtaining of a characteristic protein fingerprint of each bacterium at a given time and determined physiological condition. This specific protein profile might be used for the unequivocal detection of pathogens in food matrices. For example, this technique was effective for rapid identification of pathogens of the Enterobacteriaceae family such as *Escherichia coli*, isolated from both milk and dairy products (cottage cheese and butter).

Analysis of the metabolite profile generated by microorganisms can be another effective way to confirm their presence in food. Specifically, it is possible to identify pathogen-specific biomarkers by untargeted metabolomics after using GC-MS for the characterization of *Escherichia coli*, *Listeria monocytogenes*, and *Salmonella enterica* bacteria. The application of the technique in a nutritious liquid medium containing spiked meat samples resulted in the finding of putative biomarkers related to these bacteria, highlighting the identification of sugars, fatty acids, amino acids, nucleosides, and organic acids.

These new analytical approaches have the ability to significantly reduce the time of analysis and produce results in a short space of time compared to conventional microbiological methods that can take up to a week to confirm the presence of pathogens. This is especially relevant in dairy industries with large processing capacity that output a significant number of products. The foodomic approach can help in the detection of mycotoxins in food. These toxins of fungal origin are secondary metabolites that may cause serious health problems once ingested, since they have been reported to possess carcinogenic, immunosuppressive, hepatotoxic, nephrotoxic, and neurotoxic effects. Dairy products figure as a major source of distribution for some of these toxins, especially aflatoxins, but also fumonisin, ochratoxin A, trichothecenes, zearalenone, T-2 toxin, and deoxynivalenol. Specifically, an upward risk of finding these compounds in cheese products has been reported as a consequence of the metabolism of producing fungal species. Hence, it is important to find new tools that enhance their detention at any point in the food chain. The omic-based approach has shown to be useful in the mycotoxic analysis of milk from farm animals. In the same way, it might also be applicable to already processed products such as cheeses.

New Words

excessive [ɪk'sesɪv]	*adj.*	过度的；过多的
allergy ['ælədʒɪ]	*n.*	过敏反应；过敏症；厌恶；反感
mildew ['mɪldju:]	*n.*	霉；霉病
	vt.	使发霉
	vi.	发霉；生霉

high performance liquid chromatography (HPLC)		高效液相色谱法
gas chromatography (GC)		气相色谱法
gas chromatography-mass spectrometry (GC-MC)		气相色谱-质谱联用
sensitivity [ˌsensə'tɪvətɪ]	*n.*	感知；觉察；灵敏度；敏感
metabolomics [mɪtæˌbə'lɒmɪks]	*n.*	代谢组学
molecule ['mɒlɪkjuːl]	*n.*	［化学］分子；微粒；［化学］摩尔
liquid chromatography (LC)		液相色谱
legislation [ˌledʒɪs'leɪʃn]	*n.*	法规；法律；立法；制订法律
spoilage ['spɒɪlɪdʒ]	*n.*	损坏；糟蹋；掠夺；损坏物
mass spectrometry (MS)		质谱分析
high resolution mass spectrometry (HRMS)		高分辨率质谱
monitored ['mɒnɪtəd]	*adj.*	监视的；监控的 动词 monitor 的过去式和过去分词
pesticide ['pestɪsaɪd]	*n.*	杀虫剂；农药
metabolite [me'tæbəˌlaɪt]	*n.*	［生化］代谢物
fingerprint ['fɪŋgəprɪnt]	*n.*	指纹
	v.	取…的指印
profiling ['prəʊfaɪlɪŋ]	*n.*	资料收集；剖析研究 动词 profile 的现在分词形式
partial least square discriminant analysis (PLS-DA)		偏最小二乘判别分析
discriminant [dɪ'skrɪmɪnənt]	*n.*	可资辨别的因素
nuclear magnetic resonance (NMR)		核磁共振
spectroscopy [spek'trɒskəpɪ]	*n.*	［光］光谱学
quantification [ˌkwɒntɪfɪ'keɪʃn]	*n.*	定量；量化
aptamer ['æptəmər]	*n.*	适体；适配子
enterotoxin [ˌentərəʊ'tɒksɪn]	*n.*	［基医］肠毒素
nanomaterial [nænəʊmæ'tiəriəl]	*n.*	纳米材料
purification [ˌpjʊərɪfɪ'keɪʃn]	*n.*	净化；提纯；涤罪
nutritional [nju'trɪʃənəl]	*adj.*	营养的
transcriptomics [trænsk'rɪptəʊmɪks]	*n.*	转录组学
ecosystem ['iːkəʊsɪstəm]	*n.*	生态系统
unequivocal [ˌʌnɪ'kwɪvəkl]	*adj.*	明确的；不含糊的

ionization [ˌaɪənaɪ'zeɪʃn]	*n.*	［化学］离子化
biomarker ['baɪɒmɑːkər]	*n.*	生物标志物
immunosuppressive [ɪ'mjuːnəʊsə'presɪv]	*n.*	免疫抑制的药
	adj.	免疫抑制力的
fatty acids	*n.*	脂肪酸
amino acid [əˌmiːnəʊ'æsɪd]	*n.*	氨基酸

Notes

1) At present, the detection of target substances in food is mainly carried out via high performance liquid chromatography (HPLC), gas chromatography (GC), gas chromatography-mass spectrometry (GC-MC), and other methods.
参考译文：目前，食品中目标物质的检测主要通过高效液相色谱法（HPLC）、气相色谱法（GC）、气相色谱-质谱联用技术（GC-MC）等方法进行。
2) Together with genomics, transcriptomics and proteomics, metabolomics is involved in the study of food omics approaches among others.
参考译文：代谢组学与基因组学、转录组学和蛋白质组学一起构成了研究食品组学的方法。
3) Therefore, metabolomics approach is a suitable solution to ensure food safety, which has become a major issue worldwide, and contaminants (i.e., organic pollutants) or toxins should be monitored in order to satisfy the consumer demand.
参考译文：因此，代谢组学是确保食品安全（已成为全球的主要问题）的合适的解决方法，同时应当监测食品中污染物（如有机污染物）或毒素，从而满足消费者的需求。
4) In contrast, untargeted analysis focuses on the detection of as many groups of metabolites as possible to obtain patterns or fingerprints without identifying nor quantifying specific compounds.
参考译文：相比之下，非靶向分析侧重于检测尽可能多的代谢物，以获得图案或指纹，而无须识别或量化特定化合物。
5) This approach utilizes chemometric tools as principal component analysis (PCA) or partial least square discriminant analysis (PLS-DA) to classify the components obtained in the samples in different regions to obtain a typical fingerprint of the sample.
参考译文：这种方法利用化学计量工具，如主成分分析（PCA）或偏最小二乘判别分析（PLS-DA），对从不同区域的样品中获得的组分进行分类，以获得样品的典型指纹。
6) For food safety detection, compared with natural enzymes, nanozymes are relatively simple in terms of their preparation and purification, not requiring complex equipment or instruments.

参考译文：对于食品安全检测，与天然酶相比，纳米酶在制备和纯化方面相对简单，不需要复杂的设备或仪器。

7) It is possible to understand microbial behavior in different exogenous conditions through the analysis of RNA molecules, since the food matrix plays an important role in the expression of genes involved in activities, such as growth, survival, or level of virulence towards the host.

   参考译文：通过分析 RNA 分子可以了解不同外源条件下的微生物行为，因为食物基质在参与生命活动（如生长、存活或对宿主的毒力水平）的基因表达中起重要作用。

8) On the other hand, the low pH caused by the presence of the endogenous bacterium was one of the factors behind the down regulation of the accessory gene regulator system, a key regulator of bacterial virulence.

   参考译文：另一方面，由内源性细菌引起的低 pH 值是附属基因调节系统下调的因素之一，该系统是细菌毒力的关键调控因子。

9) The technique matrix-assisted laser desorption/ionization and subsequent analysis of the ions produced by time-of-flight (MALDI-TOF) seems to have fulfilled the expectations of the researches, providing satisfactory results.

   参考译文：基质辅助激光解吸/电离技术以及飞行时间（MALDI-TOF）产生的离子的后续分析似乎满足了研究的预期，提供了令人满意的结果。

10) For example, this technique was effective for rapid identification of pathogens of the Enterobacteriaceae family such as *Escherichia coli*, isolated from both milk and dairy products (cottage cheese and butter).

    参考译文：例如，该技术可有效快速地鉴定肠杆菌科病原菌，如从牛奶和乳制品（白软干酪和黄油）中分离的大肠埃希菌。

11) Fingerprinting is a fast, convenient and effective tool to classify samples based on metabolite patterns and it identifies the most important regions of the spectrum for further analysis.

    参考译文：指纹图谱是一种基于代谢物模式对样品进行分类的快速、方便和有效的工具，它可以识别光谱中最重要的区域以供进一步分析。

12) The food omics approach can help in the detection of mycotoxins in food. These toxins of fungal origin are secondary metabolites that may cause serious health problems once ingested, since they have been reported to possess carcinogenic, immune suppressive, hepatotoxic, nephrotoxic, and neurotoxic effects.

    参考译文：食品组学方法有助于检测食物中的霉菌毒素。这些真菌毒素是次级代谢产物，一旦摄入，可能会导致严重的健康问题。据报道，这些真菌毒素具有致癌、免疫抑制、肝毒性、肾毒性和神经毒性作用。

13) These new analytical approaches have the ability to significantly reduce the time of analysis and produce results in a short space of time compared to conventional microbiological methods that can take up to a week to confirm the presence of pathogens.

参考译文：传统的微生物方法需耗时一周来确定病原菌存在情况。与传统微生物方法相比，这些新的分析方法能够显著减少分析时间，并在短时间内产生结果。

14) Dairy products figure as a major source of distribution for some of these toxins, especially aflatoxins, but also fumonisin, ochratoxin A, trichothecenes, zearalenone, T-2 toxin, and deoxynivalenol.

参考译文：乳制品是某些毒素的主要分布来源，特别是黄曲霉毒素，但也包括伏马菌素、赭曲霉毒素 A、三氯乙烯、玉米赤霉烯酮、T-2 毒素和脱氧雪腐镰刀菌烯醇。

Lesson 3 HACCP and Food Safety

Food safety is concerned with the protection of food products from unintentional contamination by an agent reasonably likely to occur in the food supply. An example of food safety is controls to prevent contamination of ground beef by *E. coli* O157: H7 and non-EHEC (enterohemorrhagic *E. coli*). Food safety is a key responsibility of the food industry irrespective of whether the food is sourced nationally or from foreign countries. Failure to maintain food safety results in recalls that cost the industry millions of dollars in lost revenues, shatters consumer confidence, destroys brands, and ultimately places unprecedented cost on the economy. Foods defense is concerned with the protection of food from intentional contamination by biological, chemical, physical, or radiological agents that are not reasonably likely to occur in the food supply.

Hazard Analysis Critical Control Point (HACCP) is a system which provides the framework for monitoring the total food system, from harvesting to consumption, to reduce the risk of foodborne illness. The system is designed to identify and control potential problems before they occur. In its Model Food Code, the Food and Drug Administration has recommended the HACCP system "because it is a system of preventive controls that is the most effective and efficient way to assure that food products are safe" (1999 FDA Model Food Code). The application of HACCP is based on technical and scientific principles that assure safe food. The food industry, including food service, supports the use of HACCP and its principles as the best system currently available to reduce and prevent foodborne illness.

Brief Background of HACCP

Since the Federal Meat Inspection Act of 1906 and the Poultry Product Inspection Act of 1957, meat and poultry manufacturers have been directly charged with producing safe, wholesome, and unadulterated foods. As a whole, the U.S. food industry has done an excellent job of maintaining the safety of its products. Unfortunately, over the years, foodborne illnesses have resulted in thousands of sicknesses and many deaths. Investigators have always tried to trace foodborne illness outbreaks back to the food processor, determine what went wrong, and then require them to fix the problem. Over time, it became clear that taking actions after the fact were inadequate to protect consumers from foodborne illness. In order to response to consumer concerns, the U.S. Department of Agriculture and the Food and Drug Administration have required plants to implement a food safety program called HACCP. This program was developed by the Pillsbury Co. for NASA in the late 1960s to ensure the complete safety of food for space missions. The program provides a proactive approach to food safety by requiring plants to identify all potential avenues of contamination (hazards) and determine how to control, reduce, and/or prevent them to ensure that safe food will be produced.

HACCP

The Food Standards Agency has developed a range of food safety management packs for different sectors of the food industry to help food business operators manage their food safety management procedures. FDA endorses the voluntary implementation of food safety management systems in retail and food service establishments. Combined with good basic sanitation, a solid employee training program, and other prerequisite programs, HACCP can provide you and your employees a complete food safety management system.

HACCP has been defined by the WHO as "a scientific, rational and systematic approach for the identification, assessment and control of hazards." Originally developed to focus on food safety hazards, the HACCP system has been successfully applied to other applications and other industries. The intent of HACCP is to help prevent known hazards and to reduce the risks that they will occur at any point in a process through the execution of seven core action:

Conduct a Hazard Analysis

The purpose of a hazardous analysis is to develop a list of hazards which are likely to cause injury or illness if they are not controlled. Points to be considered in this analysis can include: skill level of employees; transport of food; serving elderly, sick, very young children, immune-compromised; volume cooling; thawing of potentially hazardous foods; high degree of food handling and contact; adequacy of preparation and holding equipment available; storage, and method of preparation. The next step is to determine if the factors may influence the likely occurrence and severity of the hazard being controlled. Finally, the hazards associated with each step in the flow of food should be listed along with the measures necessary to control the hazard.

Determine the Critical Control Points

A critical control point (CCP) is a point, step, or procedure where the hazard that's associated with the food can be prevented, eliminated, or reduced to acceptable levels. Some

CCPs can include cooking. For example, raw poultry is associated with bacteria. The bacteria can be eliminated, reduced, or prevented by cooking raw poultry to 165℉ for 15 s. As another example, reheating can be a CCP. If we take a stew for example, a beef stew if we reheat that beef stew to 165℉ for 15 s, we can eliminate any bacteria that might have grown during the cooling and the cold holding steps. Holding can also be considered a CCP. Cold holding, hot holding, or using time control during holding can prevent bacterial growth. It is important to consider, whether using cooking, cooling, holding, or reheating, time is a factor in any one of these CCPs.

Establish Target Levels and Critical Limits

A critical limit ensures that a biological, chemical or physical hazard is controlled by a CCP. Each CCP should have at least one critical limit. Critical limits must be something that can be monitored by measurement or observation. They must be scientifically and/or regulatory based. Examples include temperature, time, pH, water activity or available chlorine.

Establish System(s) to Monitoring CCPs

Monitoring is the scheduled measurement or observation of a CCP relative to its critical limits. The monitoring procedures must be able to detect loss of control at the CCP. Further, monitoring should ideally provide this information in time to make adjustment and ensure control of the process to prevent violating the critical limits. Where possible, process adjustments should be made when monitoring results indicate a trend towards loss of control at a CCP. The adjustments should be taken before a deviation occurs. Data derived from monitoring must be evaluated by a designated person with knowledge and authority to carry out corrective actions when indicated. If monitoring is not continuous, then the amount or frequency of monitoring must be sufficient to guarantee the CCP is in control. Most monitoring procedures for CCPs will need to be done rapidly because they relate to on-line processes and there will not be time for lengthy analytical testing. Physical and chemical measurements are often preferred to microbiological testing because they may be done rapidly and can often indicate the microbiological control of the product. All records and documents associated with monitoring CCPs must be signed by the person(s) doing the monitoring and by a responsible reviewing official(s) of the company.

Establish an Appropriate Corrective Action Plan for Each CCP

Specific corrective actions must be developed for each CCP in the HACCP system in order to deal with deviations when they occur. The actions must ensure that the CCP has been brought under control. Actions taken must also include proper disposition of the affected product. Deviation and product disposition procedures must be documented in the HACCP record keeping.

Establish Procedures to Verify that the HACCP System is Working Effectively

Every HACCP plan should be examined to validate its ability to control food safety hazards that are reasonably likely to occur, and to show that the plan is being effectively implemented. Verification should include, at a minimum: review of the HACCP system and its records; review of deviations and product dispositions; confirmation that CCPs are kept under control.

Establish Documentation Concerning All Procedures and Keep Records of their Application

The HACCP plan must be on file at the handling facility. It must include documentation relating to CCPs and any action taken on deviations or disposition of product.

Types of records could include:

a. Processing: records of all monitored CCPs.

b. Deviation: records of any deviations from the HACCP plan.

c. Ingredients: supplier qualification, ingredient certification and audit records.

d. Product safety: records on safe shelf-life, microbiological testing and microbiological challenge studies.

e. Storage and distribution: traceability data.

f. Validation studies.

Because the approach enforces procedural governance and rigorous documentation practices, HACCP serves not only as a model to assess risk, but also as an effective means to communicate risk control. This document presents some core steps in the execution of HACCP. Successful application of any risk management model requires that tools are used in concert with the quality risk management process. This guide will present the seven principles of HACCP in the context of the ICH Q9 defined quality risk management process consisting of Risk Assessment, Risk Control, Risk Review and Communication.

The HACCP system is focused on safety hazards-typical safety hazards include biological, chemical or physical agents or operations that could lead to illness or injury if not controlled typically, includes an entire process (e.g., for pharmaceutical manufacturing, it typically encompasses raw material receipt through distribution to the consumer). It is implemented through a plan that details the initial analysis and the on-going monitoring and review. The plan is a "living" system, not a one-time assessment. Generates a "HACCP Control Chart" listing the potential hazards, preventive measures, critical limits, monitoring systems and corrective actions associated with each CCP.

HACCP is a tool to assess hazards and establish control systems that focus on prevention rather than relying mainly on end-product testing. Any HACCP system is capable of accommodating change, such as advances in equipment design, processing procedures or technological developments. HACCP could be applied throughout the food chain from the primary producer to final consumer and its implementation should be guided by scientific evidence of risks to human health. As well as enhancing food safety, implementation of HACCP can provide other significant benefits. In addition, the application of HACCP systems can aid inspection by regulatory authorities and promote international trade by increasing confidence in food safety.

HACCP is the final stage of an integrated, proactive food safety program targeting the handler and designed to prevent contamination before it occurs. For HACCP to be a fully functional part of overall product safety management, well-established, comprehensive prerequisite programs must be in place; an integrated HACCP program is of no benefit without such programs. Good Agricultural Practices (GAPs) provide guidelines to growers on how to minimize potential biological hazards during production and harvesting of food materials. GMPs define procedures to be used by handlers to allow food materials to be processed, packed and sold under sanitary conditions. SSOPs ensure a clean and sanitary environment in the facility. Together, these programs provide a framework for an HACCP program by proactively eliminating or minimizing potential sources of contamination. HACCP provides a systematic approach to identify, assess and control the risk of biological, chemical and physical hazards that can be reduced, prevented or eliminated. The idea is to develop a plan that anticipates and identifies places in the production process-known as CCPs—where contaminants might be introduced or other food safety concerns can be identified. When critical limits are exceeded, corrective action must be taken and documented. An independent third party should be used to verify or validate the effectiveness of an HACCP plan.

New Words

agrochemicals [æg'rɒkemɪklz] *n.* 农业化学试剂

chronic ['krɒnɪk] *adj.* 长期的；慢性的；惯常的

unconventional [ˌʌnkən'venʃənl] *adj.* 非传统的；不合惯例的；非常规的

melamine ['meləmiːn] *n.* 三聚氰胺

retardant [rɪ'taːdənt] *n.* 阻滞剂；抑止剂
adj. 延缓的；阻碍的

glazed [gleɪzd] *adj.* 像玻璃的；目光呆滞的；光滑的
动词 glaze 的过去式和过去分词

surfactant [sɜː'fæktənt] *n.* 表面活性剂；表面活化物质
adj. 表面活化剂的

caulking ['kɔːkɪŋ] *n.* 堵塞材料

seedling ['siːdlɪŋ]	*n.*	幼苗
terrestrial [tə'restriəl]	*adj.*	陆地的；陆生的；地球的
	n.	地球生物
histamine ['hɪstəmiːn]	*n.*	组织胺
veterinary ['vetnri]	*adj.*	兽医的
	n.	兽医
genera ['dʒenərə]	*n.*	种类；属；genus 的复数形式
cyanide ['saɪənaɪd]	*n.*	氰化物
	v.	用氰化物处理
lethality [lɪ'θælɪtɪ]	*n.*	致命性；杀伤力

Notes

1) Finally, the hazards associated with each step in the flow of food should be listed along with the measures necessary to control the hazard.
   参考译文：最后，应该列出食品生产线中与每一个操作步骤有关的危害以及控制此危害的必要措施。
2) Every HACCP plan should be examined to validate its ability to control food safety hazards that are reasonably likely to occur, and to show that the plan is being effectively implemented.
   参考译文：对于每一个 HACCP 计划，都应验证其控制所产生的食品安全危害效果，从而显示 HACCP 计划能够被有效地实施。
3) For HACCP to be a fully functional part of overall product safety management, well-established, comprehensive prerequisite programs must be in place; an integrated HACCP program is of no benefit without such programs.
   参考译文：作为整个产品安全管理的关键部分的 HACCP 计划，既定的全面而必要的方案必须到位；若没有这些方案，只有一个完整的 HACCP 计划是没有任何意义的。

Reading Material 1 Food Poisoning

We are now in the 21st century, food safety issues have as high a priority and significance as they did over 100 years ago. Public concerns have arisen regarding high-profile food-borne illness outbreaks due to contamination of food with certain pathogens (e.g., *Salmonella*, *Escherichia coli* O157:H7, *Listeria monocytogenes*, and others) which have serious acute impact and potential chronic long-term complications in the ever-increasing high-risk population segment (e.g., elderly, children, immuno-compromised people). In addition, food-borne illness outbreaks are occurring in foods previously not considered high risk (e.g., fruit juices, fresh produce, and deli meats).

Foodborne illness, or food poisoning, occurs commonly throughout the world. While the American food supply is among the safest in the world, the federal government estimates that there are about 48 million cases of foodborne illness annually—the equivalent of sickening 1 in 6 Americans each year. Each year these illnesses result in an estimated 128,000 hospitalizations and 3,000 deaths. They also cause billions of dollars in healthcare-related industry costs annually, which means that the average person in the US will have a foodborne illness once every three to four years. Fortunately, most people recover from an episode of foodborne illness without any need to consult with a healthcare provider or any long-term complications.

Foodborne illness (also foodborne disease and colloquially referred to as food poisoning) is any illness resulting from the consumption of contaminated food by pathogenic bacteria, viruses, or parasites, as well as chemical or natural toxins such as poisonous mushrooms. These contaminants can get into food in many different ways. The most common causes of contamination are from infected food handlers who don't wash their hands or wear gloves properly and from kitchenware or service utensils that aren't cleaned properly. Food poisoning can also occur when foods haven't been refrigerated at cold enough temperatures or cooked at hot enough temperatures.

Some of the most common factors that cause food poisoning include the following:

Chemicals

Food chemical safety remains a serious concern to the food industry. Risks such as adulteration, the existence of toxic and allergenic compounds in foods, and poor regulation of postharvest processing, indicate that food chemical safety is not fully guaranteed.

Considerable advances have been made over the past century in the understanding of the chemical hazards in food and ways for assessing and managing these risks. At the turn of the 20th century, many Americans were exposed to foods adulterated with toxic compounds. In the 1920s the increasing use of insecticides led to concerns of chronic ingestion of heavy metals such as lead and arsenic from residues on crops. By the 1930s, a variety of agrochemicals were commonly used, and food additives were becoming common in processed foods. Between the 1940s and 1950s, advances were made in toxicology, and more systematic approaches were adopted for evaluating the safety of chemical contaminants in food. Modern GC and liquid chromatography, both invented in the 1950s and 1960s, were responsible for progress in detecting, quantifying, and assessing the risk of food contaminants and adulterants. In recent decades, chemical food safety issues have been the center of media attention include the presence of natural toxins, processing-produced toxins, heterocyclic aromatic amines, and furan(=furane, *n.* 呋喃), food allergens, heavy metals, industrial chemicals, contaminants from packaging materials, and unconventional contaminants(melamine, *n.* 三聚氰胺) in food and feed. Due to the global nature of the food supply and advances in analytic capabilities, chemical contaminants will continue to be an area of concern for regulatory agencies, the food industry, and consumers in the future.

Environmental Contaminants

Food represents a part of the global environment which can be contaminated by chemicals from many different sources. Following their release into the environment (soil, sediment, water, air), chemical contaminants can enter plants and animals at the bottom of the food chain, which are then consumed by animals higher up. The chemicals contained in these animals and plants can enter our bodies when we consume them as food (e.g., meat, dairy products, fish, vegetables, fruit). Contamination route in food chain is especially important for chemicals persisting and accumulating in the environment, such as lead, dichloro-diphenyl-trichloroethane (DDT, *n.* 二氯二苯三氯乙烷), polychlorinated biphenyls (PCBs, *n.* 多氯联苯) and brominated flame retardants, which is also relevant for chemicals that are used in large amounts and occurs ubiquitously in the environment, for example, phthalates (*n.* 邻苯二甲酸酯) packaging, perfluorinated (*adj.* 全氟化的) chemicals used in grease proof packaging for fast foods. It is for these reasons that WWF (World Wildlife Fund)'s focus now shifts to investigating contaminants in food, and in the process by its human biomonitoring work.

Lead is a naturally occurring metal found in rock and soil, also has many industrial applications. Due to both its natural occurrence and long history of global use, lead is ubiquitous in the environment, which presents in air, water and soil as well as in food, drinking water and household dust. Levels of lead in most environmental media have declined significantly over the past few decades due to the discontinued use of lead in paint, gasoline and the solder used in food cans. After the phase-out of leaded gasoline and the subsequent reduction of airborne lead, food and drinking water are the primary sources of lead exposure to adults within the general population. In addition to food and drinking water, the ingestion of house dust and soil containing lead can also significantly contribute to the lead exposure of infants and toddlers. Lead is not permitted to be added to foods sold in many countries. However, due to its ubiquitous presence in the environment, it is present in all foods, generally at very low levels. Lead can enter the food chain through various pathways. For example, plants uptake lead from the soil and airborne lead may also be deposited on their surfaces. Also, fish can absorb lead from water and sediments. While other animals may be exposed to lead through the foods they eat. Lead may also be introduced to foods from the use of lead containing dishware such as lead glazed pottery or lead crystalware. The preparation of foods with water containing lead can also introduce lead to foods. Consuming wild game that has been shot with lead bullets is another potential source of dietary lead exposure.

Perfluorinated chemicals (PFCs, 全氟化合物) are man-made chemicals that are used in commercial and consumer products and also have a variety of industrial applications. Perfluorooctane sulfonate (PFOS, 全氟辛烷磺酸) is perhaps the most well-known PFC and has been used as, among other things, water, stain, and oil repellent for textiles, carpet, and food packaging, a surfactant in the electroplating industry, and an additive in fire-fighting foams. Due to their persistence and widespread use, PFCs have been detected at low concentrations in the environment, food, and in human blood in several different countries.

PCBs are a group of persistent chemicals used in electrical transformers and capacitors for insulating purposes, in gas pipeline systems as a lubricant, and in caulks and other building materials. The sale and new use of PCBs were banned by law in 1979, although continued use in existing electrical equipment is allowed, but large reservoirs of PCBs still remain in the environment. PCBs easily accumulate in fat tissue, so they are commonly found in foods derived from animals. Consumption of fish is an important source of PCB exposure, but other foods with lower PCB levels that are consumed more frequently, including meat and poultry, are also important contributors to PCB exposure. Some research has found that dairy products can contribute to PCB exposure. Exposure to PCBs remains widespread；however, declining environmental levels of PCBs mean that children exposed to PCBs through their diet today are being exposed to lower levels compared with children in previous generations.

Natural Toxins

As opposed to man-made chemicals such as pesticides, veterinary drugs or environmental pollutants that get into our food supply, and natural toxins can be present due to their natural occurrence in food. Natural toxins are chemicals that are naturally produced by living organisms. These toxins are not harmful to the organisms themselves but they may be toxic to other creatures, including humans, when eaten.

Toxic compounds are produced by a variety of plants and animals. Natural toxins may be present serving specific function in the plant and animal or evolved as chemical defense against predators, insects or microorganisms. These chemicals have diverse chemical structures and are vastly different in nature and toxicity. Natural toxins found inherently in foods of plant and animal origins can be harmful when consumed in sufficient quantities.

Pesticides

Pesticides refer to a broad class of crop-protection chemicals: insecticides, which are used to control insects；rodenticides, which are used to control rodents；herbicides, which are used to control weeds；and fungicides, which are used to control fungi, mold and mildew. Herbicides are the most widely used chemicals in agriculture. Pesticides help control hundreds of weed species, more than one million species of harmful insects and some 1,500 plant diseases. Pest problems and their management vary widely throughout the country based on climate, soil types and many other conditions. As a result, chemical pest control has won a central place in modern agriculture, contributing to the dramatic increases in crop yields achieved in recent decades for most major field, fruit and vegetable crops. The use of pesticides, it helps growers to produce some crops profitably in otherwise unsuitable locations, extend growing seasons, maintain product quality and extend shelf life.

Viruses and Bacteria

Bacteria and viruses are the most common cause of food poisoning. The symptoms and severities of food poisoning vary, depending on which bacteria or virus has contaminated the food. In the US, about 31 viral and bacterial pathogens are responsible for almost 9.4 million diagnosed food poisoning illnesses per year. About 39 million food poisoning cases are unspecified (undiagnosed). The most common pathogens that cause food poisoning are *Norovirus*, *Salmonella*, *Staphylococcus aureus*, *Escherichia coli* and so on.

Physical Contaminants

Physical contaminants are substances that become part of a food mixture. They may not change or damage the food itself. However, their presence can create health hazards for the consumer. For instance, metal filings or broken pieces of glass have occasionally gotten into foods. These materials would not spoil food, but they could cause injury if swallowed. Other physical contaminants include packaging material, insects, and rodent droppings. Insect and rodent contamination present two major problems. The first is the large volume of food that insects or rodents can eat and/or destroy. It is estimated that as much as 10% of the U.S. grain crop is destroyed annually by insects. The second concern is the microbes that may enter the food because of the insects or rodents. For example, flies pick up microbes on their hairy feet. When flies walk on food, microbes can transfer from their feet to the food. Insects and rodents also damage the surfaces of foods such as fruits and vegetables. This creates openings that allow microbes to enter and multiply within the foods. Insects and rodents can contaminate the food supply at any stage of growth or production. For example, some insects lay eggs in wheat while it is growing in fields. Their eggs are not visible, and presence in small amounts is not harmful to human health. Keeping all insect eggs out of the wheat supply would be extremely expensive. On the other hand, cockroaches are likely to enter the wheat supply during the processing stage. Their presence is less acceptable and can be affordably controlled by the food manufacturer.

New Words

revenue ['revənjuː]	*n.*	税收；国家的收入；收益
shatter ['ʃætə (r)]	*v.*	粉碎；打碎；破坏；破掉

	n.	碎片；乱七八糟的状态
poultry ['pəʊltri]	*n.*	家禽；家禽肉
wholesome ['həʊlsəm]	*adj.*	健全的；有益健康的；合乎卫生的；审慎的
retail ['riːteɪl]	*v.*	零售；转述
	n.	零售
	adv.	以零售方式
	adj.	零售的
scheduled ['ʃedjuːld]	*adj.*	预定的；已排定的
anticipate [æn'tɪsɪpeɪt]	*vt.*	预期；提前使用；抢…前
	vi.	预言
implementation [ˌɪmplɪmen'teɪʃn]	*n.*	履行；落实；装置
disposition [ˌdɪspə'zɪʃn]	*n.*	性情；倾向；安排；处置；控制

Notes

1) Foodborne illness (also foodborne disease and colloquially referred to as food poisoning) is any illness resulting from the consumption of contaminated food by pathogenic bacteria, viruses, or parasites, as well as chemical or natural toxins such as poisonous mushrooms.
参考译文：食物传染疾病（也称食源性疾病，俗称食物中毒）是食用了被致病细菌、病毒或寄生虫，以及化学药品或天然毒素（如蘑菇）所污染的食品而导致的一类疾病。

2) Natural toxins may be present serving specific function in the plant and animal or evolved as chemical defense against predators, insects or microorganisms.
参考译文：在植物和动物中存在的天然毒素可能具有特定的功能，或者作为化学防御剂来抵御捕食者、昆虫或微生物。

Exercise

1. Translate the following sentences into Chinese.

1) Physical contaminants are substances that become part of a food mixture. They may not change or damage the food itself. However, their presence can create health hazards for the consumer.

2) Bacteria and viruses are the most common cause of food poisoning. The symptoms and severities of food poisoning vary, depending on which bacteria or virus has contaminated the food.

2. Translate the following sentences into English.

1）当食物没有在足够冷的温度下冷藏或在足够热的温度下煮熟时，也会引起食物中毒的发生。

2）食品的化学安全仍然是食品行业严重关切的问题。掺假、食品中存在有毒和致敏化合物、采后加工监管不力等风险表明，食品的化学安全没有得到充分保障。

Reading Material 2 Safety of Pickled Foods

Pickling is the process of preserving food by anaerobic fermentation in brine or vinegar. The resulting food is called a pickle. This procedure gives the food an interesting twist in flavor. Pickled foods are a category of food products that is very popular with food entrepreneurs and farmers interested in value. Many products are traditionally processed by this way, including pickles (cucumbers) and pickled vegetables, meat and eggs. The variety of products and flavors is limited only by the creativity of food processors as new formulations and presentations continuously debut in food stores to meet consumer's new expectations and ethnic preferences.

There are two classifications of pickles, fresh pack and fermented.

Fresh Pack Pickles

Fresh pack pickles, such as fresh-pack dill (小茴香) pickles, crosscut pickles slices (bread-and-butter pickles), sweet gherkins (腌食用小黄瓜), or cauliflower pickles are quick and easy to prepare. Some recipes call for soaking vegetables for a few hours or overnight in brine (salt water). After soaking, they are drained, packed into clean, hot jars, and covered with boiling-hot pickling solution containing vinegar and spices. They are not fermented. The vinegar, that should be 5% acetic acid, acts as the preservative. The short brining procedure serves two purposes. First, it removes bitter juices present in some cucumbers and enhances the uptake of pickling solution, resulting in a firmer product. Too strong of a brine shrivels the cucumbers. Too weak of a brine removes too little water, which results in spoilage. Processing in a boiling water canner is necessary for both fermented and fresh-pack pickles.

The jars are then closed with two-piece canning lids and processed in a boiling water bath. The pickling solution must contain plenty of vinegar. Some old recipes call for too little vinegar to ensure safety. Never decrease the vinegar in a pickling recipe. Avoid packing vegetables too tightly into jars. Plenty of room must be left for the boiling-hot pickling solution to circulate around pickles during processing. Spoilage is a problem when pickles are wedged tightly into jars.

Fermented Pickles

Fermented pickles also called crock pickles (lacto-fermented or naturally fermented). It is the original pickling method and has been an essential part of healthy human diets throughout the world for thousands of years. In Asia, consumption of fermented vegetables dates back at least to the third Century B.C., when the Great Wall of China was being constructed. In Europe, sauerkraut is known to have been an important food among the ancient Romans.

Fermented pickles require more time and effort to make than fresh-pack pickles. Brined dill pickles are an example of fermented pickles that are prepared by soaking cucumbers in brine (salt water) for about 3 weeks. During this time, lactic acid bacteria that are naturally present on vegetables convert carbohydrates (sugars) in the vegetables into lactic acid by a process known as fermentation. Lactic acid not only preserves the pickles but also gives them good flavor. (Vinegar, that contains acetic acid, gives fresh-pack pickles a "sharper" flavor.)

When making the brine, measure the salt and water carefully. It is important to get just the right concentration of salt so the lactic acid bacteria that can tolerate salt and will be able to grow. Most spoilage organisms cannot tolerate salt and will die in the brine. If the brine is too salty, even the lactic acid bacteria will die. If the brine is not salty enough, undesirable organisms will grow and spoil the pickles. Make the brine with cold or room temperature water. Do not use boiling water it will kill the lactic acid bacteria. During fermentation, keep the pickles at room temperature between 68°F and 78°F. Fermenting pickles must be kept submerged. Uncovered pickles will spoil. Use a plate to cover the pickles and weight it down with a glass jar or plastic bags filled with brine (6 tablespoons salt to 1 gallon of water). Remove any scum that forms on the surface of the brine daily. The scum consists of yeasts that destroy lactic acid and produce enzymes that break down pectic substances in the pickles, making them soft and mushy. If the scum is not removed daily, pickles will spoil. After three or more weeks, fermentation should be complete. The pickles will have an olive-green color and a desirable flavor. The brine will be cloudy as a result of yeast growth during the fermentation period. Strain the brine and then heat it to boiling. Pack the pickles into clean, hot jars. Do not wedge tightly. Cover with boiling hot brine. Put lids on the jars and process in a boiling water bath.

This traditional pickling process went out of favor with the advent of industrial food production. Modern pickling methods, including use of vinegar (usually in place of fermentation) and pasteurization, produce a uniform, shelf stable product suitable to the needs of the large food corporations. Unfortunately, modern pickles do not offer the authentic flavor or health-promoting qualities of traditional pickles.

Safety of Pickled Foods

A distinguishing characteristic of pickled foods is a pH 4.6 or lower, which is sufficient to kill most bacteria. Pickling can preserve perishable foods for months. Antimicrobial herbs and spices, such as mustard seed, garlic, cinnamon or cloves, are often added. If the food contains sufficient moisture, a pickling brine may be produced simply by adding dry salt. For example, German sauerkraut and Korean kimchi are produced by salting the vegetables to draw out excess water. Natural fermentation at room temperature, by lactic acid bacteria, produces the required acidity. Other pickles are made by placing vegetables in vinegar. Unlike the canning process, pickling (which includes fermentation) does not require that the food be completely sterile before it is sealed. The acidity or salinity of the solution, the temperature of fermentation, and the exclusion of oxygen determine which microorganisms dominate, and determine the flavor of the product.

When both salt concentration and temperature are low, *Leuconostoc mesenteroides* (肠膜明串珠菌) dominates, producing a mix of acids, alcohol, and aroma compounds. At higher temperatures *Lactobacillus plantarum* (植物乳杆菌) dominates, which produces primarily lactic acid. Many pickles start with *Leuconostoc* (明串珠菌属), and change to *Lactobacillus* (乳杆菌属) with higher acidity.

The pH value of 4.6 is important because it is the limiting factor for the growth of an extremely dangerous microorganism called *Clostridium botulinum*(丁酸梭菌), which produces a potent toxin that causes the lethal disease botulism. The regulations concerning acidified foods were established to assure the control and inhibition of the growth of *Clostridium botulinum* by proper acidification and pH control, as this microorganism is very heat resistant and therefore it is not destroyed by pasteurization or cooking temperatures below 212°F.

The WHO has listed pickled vegetables as a possible carcinogen, and the British Journal of Cancer released an online 2009 meta-analysis of research on pickles as increasing the risks of esophageal (*adj.* 食管的、食道的) cancer. The report cites a potential two-fold increased risk of esophageal cancer associated with Asian pickled vegetable consumption. Results from the research are described as having "high heterogeneity" and the study said that further well-designed prospective studies were warranted.

Notes

1) Unfortunately, modern pickles do not offer the authentic flavor or health-promoting qualities of traditional pickles.
   参考译文：很遗憾的是，相较于传统发酵发生产的泡菜，现代的泡菜不具有独特的风味，或不具有促健康的功能。
2) The acidity or salinity of the solution, the temperature of fermentation, and theexclusion of oxygen determine which microorganisms dominate, and determine the flavor of theproduct.
   参考译文：泡菜汁的酸度或盐度、发酵温度以及厌氧的程度决定了哪类微生物占优势，进而决定了泡菜的风味。

Exercise

1. Please translate the following sentences into English.

1）食盐是泡菜加工过程中的重要成分，它主要起到改善泡菜的口感和风味以及防止泡菜腐败变质的作用。

2）酸度、盐浓度、发酵温度以及厌氧条件是决定泡菜风味以及微生物区系的四个关键因素。

2. True or false.

1) Pickling (which includes fermentation) requires that the food be completely sterile before it is sealed. ( )

2) The pH value is the limiting factor to control the growth of *Clostridium botulinum*.( )

3) Yeasts are benefit to for the texture of pickle. ( )

4) Modern pickling method, vinegar is an essential ingredient. ( )

Chapter 7

New Technique of Food Science

Lesson 1 Application of Biotechnology in Food Science

Out of the several definitions of biotechnology, perhaps one of the broadest is the use of living cells, microorganisms, or enzymes for the manufacture of chemicals, drugs, or foods or for the treatment of wastes. A simpler approach defines biotechnology as the use of biological organisms or processes in any technological application. Biotechnology has had a tremendous impact on the food industry. It has provided high-quality foods that are tasty, nutritious, convenient, and safe, and it has the potential for the production of even more nutritious, palatable, and stable food.

Traditional Biotechnology

The roots of biotechnology can be found in the ancient processes of food and beverage fermentation. These traditional technologies are present in almost every culture in the world and have evolved over many years without losing their traditional essence. Examples of these processes include the production of some well-known foods, such as bread, wine, yogurt, and cheese. These products, like many others (such as ripened sausages, pickles, sauerkraut, soy sauce, vinegar, beer, and cider), are produced using the natural processes of living organisms (e.g., fermentation)—in other words, by using biotechnology. Some relatively new developments in these traditional products include bio-yogurts, or biogurts, which contain extra bacteria (usually probiotic organisms) that are not found naturally in the original food. These probiotic bacteria most often include *Lactobacillus acidophilus* and *Bifidobacterium bifidum*. Another traditional biotechnology technique is the production of mycoprotein Quorn™ as an alternative to meat, which was developed muchmore recently. Because these techniques are considered conventional, they have not caused public concern.

Enzymes are also widely used in the food and beverage industries. For economic reasons, they are used in a relatively crude form or in a reusable form, usually achieved by immobilization. The dairy industry uses primarily rennins and lactases, the brewing industry, proteases and amylases. Highfructose corn syrup is produced from starch by using α-amylase, amyloglucosidase, and glucose isomerase, and the beverage industry consumes a great amount of pectinases.

Molecular (Modern) Biotechnology

"Modern" or molecular biotechnology, in contrast with "traditional" biotechnology, also includes the use of techniques of genetic engineering—that is, techniques for altering the properties of biological organisms. This allows characteristics to be transferred between organisms to give new combinations of genes and improved varieties of plants or microorganisms for use in agriculture and industry. Some consumers are concerned about the safety of using the techniques of modern biology, although in countries such as Japan and the U.S. consumers remain optimistic about biotechnology. They are generally willing to purchase foods developed through these techniques, and the food and agricultural applications are as acceptable as new medicines. As stated above, recombinant DNA techniques have revolutionized the fields of biology, biochemistry, and biotechnology, as they have made research of genomes more possible than ever. For example, molecular cloning and the polymerase chain reaction (PCR) have been used in order to obtain the large number of DNA copies required for DNA sequencing methods.

Molecular Cloning

In this technique, a fragment of DNA isolated from a donor cell (e.g., bacteria, yeast, or any animal or plant cell) is incorporated into a vector (e.g., plasmids or phages), by which the gene of interest can be introduced into a host cell. The formation of the recombinant DNA molecule requires a restriction endonuclease to cut and open the vector DNA. After the sticky ends of the vector have been annealed with those of the donor DNA, a DNA ligase joins the two molecules covalently. The recombinant molecule is then inserted into bacterial cells by any suitable method such as electroporation. It is necessary that the recombinant vectors contain regulatory regions recognized by the bacterial enzymes.

PCR

PCR considered one of the most significant DNA technologies developed, is a method for amplifying very small amounts of DNA that copies part of a genome for subsequent sequencing or other analyses. It is an *in vitro* laboratory method to amplify a specific segment of a genome DNA by using a pair of specific primers (oligonucleotides used to start the DNA replication) to allow a section of the DNA to be repeatedly copied. Template strands are separated by heating and then cooled to allow primers to anneal to the template. The temperature is raised again to allow primer extension (template strand copying) by means of a thermostable DNA polymerase.

The procedure is repeated 30-40 times with an exponential amplification of the DNA concentration. With the availability of this technique, enormous quantities of genes have now been sequenced for a wide range of organisms. The genomes of several bacteria and small organisms have already been fully sequenced, and the genomic sequences of many higher organisms such as plants, animals, and humans (around 90% complete) have been published.

Applications of Molecular Biotechnology

The application of biotechnology in food sciences has subsequently increased the production of food and enhanced its quality and safety. Food biotechnology is a dynamic field and the continual progress in the field has not only dealt effectively with the issues relating to food security, but also augmented the nutritional and health aspects of food. In the past decade, major breakthroughs have happened and enormous progress has been made in field of food biotechnology, including improvements in industrial process technology, farming system for growing and harvesting food, genetic improvements to organisms used in the food supply, and use of advanced techniques to monitor food safety, nutritional quality, flavor, texture and their shelf life. Food biotechnology, begins with exploring the role of microbes in food fermentation, has now progressed to increasing the shelf life of food and enhancing its flavor. Presently, there has been a shift in the focus of biotechnological progress to find out new approaches in food fermentation and develop multifunctional microorganisms to improve the nutritional and health benefits of food. Major challenges facing the world today are not just those of food production and food quality for meeting the protein and calorie needs, but also those related for better health. So implementation of any food produced through biotechnology should have to cross the environmental and ethical issues barrier. Today, it is expected from food biotechnologists that they satisfy many requirements related to health benefits, sensory properties and possible long-term effects associated with the consumption of food produced by biotechnology. Therefore, several researchers across the globe are investigating novel, biological, molecular and biochemical strategies by using cutting-edge technologies and state-of-the-art techniques for improving the food production and processing, for enhancing the safety and quality of food ingredients for better human health. The progress in food biotechnology is a ray of hope to tackle the food security of the over explosive population especially in developing countries. Following are some examples of the application of the biotechnology.

Plant Biotechnology for Food Production

Plant genetic engineering may be defined as the manipulation of plant development, structure, or composition by insertion of specific DNA sequences. These sequences may be derived from the same species or even variety of plant. This may be done with the aim of altering the levels or patterns of expression of specific endogenous genes—in other words, to make them more or less active or to alter when and where in the plant they are switched on or off. Alternatively, the aim may be to change the biological (e.g., regulatory or catalytic) properties of the proteins they encode. However, in many cases, genes are derived from other species, which may be plants, animals, or microbes, and the objective is to introduce novel biological properties or activities.

To date, numerous transgenic plants have been generated, including many crop and forest species. In the near future, plant biotechnology will have an enormous impact on conventional breeding programs, because it can significantly decrease 10 to 15 years that it currently takes to develop a new variety using traditional techniques；further, it will also be used to create plants with novel traits.

From a biotechnology point of view, there exist two areas in which plant genetic engineering is being applied as a means of enhancing the rational exploitation of plants. In the first, the addition of genes often improves the agronomic performance or quality of traditional crops. Thus, tremendous progress has occurred in the genetic engineering of crop plants for disease-, pest-, stress-, and herbicide-resistance traits, as well as for traits that enhance shelf life and processing characteristics of harvested plant materials. Some genetically determined agricultural traits, such as recombinant resistance approaches or delay of fruit senescence and ripening, are attractive because they involve only minorchanges to the plant (i.e., the introduction of a single heterologous gene). Consequently, the characteristics of commercially successful cultivars are likely to remain unmodified due to genetic improvement.

In the second major area, genetic manipulation is being exploited with the objective of improving the quality of plant products consumed by human beings, and that will affect their nutrition and health. This has already yielded more nutritious grains with modified oil, protein, carbohydrate content, and composition；process-improved flours；designer oilseeds with tailored end-uses；and plants producing high-value biomolecules, such as milk and pharmaceutical proteins, industrial enzymes, vitamins, pigments, nutraceuticals, and edible vaccines for the food and pharmaceutical industries. Among the approaches used to produce such transgenic plants with new quality attributes, which represent the second generation of genetically modified plants, are (a) manipulation of plant endogenous metabolic pathways in order to favor the accumulation of important and desired products, and (b) generation of transgenic plants that can act as living bioreactors or biofactories.

Improvement of Plant Nutritional and Functional Quality

Human beings require a diverse, well-balanced diet containing a complex mixture of both macronutrients and micronutrients in order to maintain optimal health. Macronutrients, carbohydrates, lipids, and proteins make up the bulk of foodstuff and are used primarily as an energy source. Modifying the nutritional composition of plant foods is an urgent worldwide health issue, as basic nutritional needs for much of the world population are still unmet. Large numbers of people in developing countries exist on diets composed mainly of a few staple foods which usually present poor food quality for some macronutrients and many essential micronutrients.

Seeds and tubers are the most important plant organs harvested by humankind, in terms of their total yield and their use for food, feed, and industrial raw material. This exploitation is possible because they contain rich reserves of storage compounds such as starch, proteins, and lipids. Genetic engineering provides an opportunity to explore and manipulate the structure, nutritional composition, and functional properties of those macromolecules to improve their food quality for traditional end uses and to introduce new properties for novel applications.

Carbohydrates

Starch is composed of two different glucan chains, amylose and amylopectin. These polymers have the same basic structure but differ in their length and degree of branching, which ultimately affects the physicochemical properties of both polysaccharides. Amylose is an essentially linear polymer of glucosyl residues joined via α-1,4 glycosidic bonds, whereas amylopectin exists as a branched α-1,4; α-1,6 D-glucan polymer. The physicochemical properties of the α-1,4 glucans are based on the extent of branching and/or polymerization.

Starch is a very important staple in the diet of the world population and is widely used in the food and beverage industries to produce glucose and fructose syrups as sweetener and to confer functional properties for food processing. Starch is primarily used as thickener, but also as binder, adhesive, gelling agent, film former, and texturizer in many snacks. The relative amounts of amylose and amylopectin are what give polysaccharides their unique physical and chemical properties, which convey specific functionality and are of biotechnological importance. Therefore, it might be of value to produce polysaccharides with features that are intermediate between amylose and amylopectin, that are more highly branched, or that have a higher molecular weight.

The basic pathways of amylose and amylopectin synthesis are well understood and genes readily cloned. Thus, the potential exists to produce plant starches with a wider range of structures and properties. The targets are varied but include mutant wheat starches to mimic those from maize, phosphorylated starch (currently obtained from potato), and resistant starches for healthy diets, among others. Most starch is synthesized from sucrose, a route that involves four steps: initiation, elongation, branching, and granule formation. In maize and many other plant species, at least 13 enzymes have been identified in the starch biosynthetic pathway. Of these, three enzymes are considered to be key in the synthesis of amylose and amylopectin:

• ADP-glucose pyrophosphorylase (AGP) is involved in the initiation step and generates the glucosyl precursor ADP-glucose from glucose-1-phosphate.

• Two distinct classes of starch synthase (involved in elongation and granule formation) are found within the plastids: those bound exclusively to the granule, known as granule-bound starch synthases (GBSSs), and others that can be found in the soluble phase or granule bound, known as starch synthases (SSs).

• Branching enzyme (involved in branching and granule formation) is a transglycosylase involved in amylopectin synthesis, also known as starch synthase of amylopectin biosynthesis.

Amylose has been shown to be the product of GBSS. The waxy mutants, which lack GBSS, have only amylopectin but still possess soluble starch synthases, suggesting that different enzymes participate in amylose and amylopectin synthesis.

New Words

Lactobacillus acidophilus		嗜酸乳杆菌
Bifidobacterium bifidum		双歧杆菌
mycoprotein [maɪkəʊ'prəʊtiːn]	*n.*	真菌蛋白；菌蛋白
pectinase [pek'tiːnəz]	*n.*	果胶酶
polymerase ['pɒlɪməreɪs]	*n.*	聚合酶
donor cell		供体细胞
endonuclease [endə'njuːklɪeɪs]	*n.*	核酸内切酶
sticky ends		黏性末端
anneal [ə'niːl]	*v.*	退火；锻炼；煅烧
DNA ligase		DNA 连接酶
electroporation [elektrəʊpɔː'reɪʃn]	*n.*	电穿孔
primer ['praɪmə (r)]	*n.*	引物
cutting-edge technologies		尖端技术
state-of-the-art techniques		最先进的技术
gelling agent		胶凝剂
plastid ['plæstɪd]	*n.*	质体；色素体；成形粒
waxy ['wæksi]	*adj.*	蜡制的；像蜡的；质地光滑的；柔软的；苍白的

Notes

1) The progress in food biotechnology is a ray of hope to tackle the food security of the over explosive population especially in developing countries.

   参考译文：食品生物技术的进步为解决爆炸性人口的食品安全问题带来了一线希望，尤其是在发展中国家。

2) This has already yielded more nutritious grains with modified oil, protein, carbohydrate content, and composition; process-improved flours; designer oilseeds with tailored end-uses; and plants producing high-value biomolecules, such as milk and pharmaceutical proteins, industrial enzymes, vitamins, pigments, nutraceuticals, and edible vaccines for the food and pharmaceutical industries.

   参考译文：已经生产出了含有改良的油脂、蛋白质、碳水化合物含量和组成的营养谷物，工艺改进的面粉，具有特定用途的改性油籽，可以用于生产牛奶和药用蛋白质、工业酶、维生素、色素、营养品等高附加值生物分子的植物，以及供食品和制药工业使用的可食用疫苗。

Lesson 2 New Sterilization Technology

Sterilization is a process, physical or chemical, that destroys or eliminates all microorganisms and bacterial spores. Conventional sterilization techniques rely on irreversible metabolic inactivation or on breakdown of vital structural components of the microorganisms. Microorganisms can be destroyed (irreversibly inactivated) by established physical microbicide treatments, such as heating, UV or ionizing radiation, by methods of new non-thermal treatments such as HHP, pulsed electric fields, oscillating magnetic fields or photodynamic effects; or a combination of physical processes such as heat-high pressure or heat-irradiation. Mechanical removal of microorganisms from food may be accomplished by membrane filtration of food liquids. As a definition for sterilization is stated: The objective of sterilization is a practical total deactivation of microorganisms and biological substances. (CEN-Norm 12740). Full canning sterilization process has been designed to achieve at least 12 log reductions of key spore-forming pathogens (mesophilic *Clostridium botulinum*) to achieve commercial sterility. By this process the product would be expected to remain microbiologically stable at ambient temperatures up to years. This lesson will introduce the characteristics and principles of various new sterilization technologies in the food industry.

Electric Heating Methods

Ohmic Heating

Ohmic (electrical resistance) heating is a heat treatment process in which an electric current is passed through the food to achieve sterilization and desired degree of cooking. Ohmic heating is alternatively called resistance heating or direct resistance heating. The current generates heat (joule effect) in the food itself, delivering thermal energy where it is needed. Ohmic heating is a high temperature short-time method (HTST) that can heat an 80% solids food product from room temperature to 129℃ in 90 seconds allowing the possibility to decrease of high temperature over processing. The control factors for commercial applications are flow rate, temperature, heating rate and holding time of the process. The factors influencing the heating in the food are the size, shape, orientation, specific heat capacity, density, thermal, electrical conductivity, and specific heat capacity of the carrier medium. In practise the ohmic method heats particulates faster than the carrier liquid (heating inversion), which is not possible by traditional, conductive heating. Although the heating rate may be uniform, the temperature distribution across the food material can vary significantly. Therefore, design of effective ohmic heaters depends on the electrical conductivity of the food. In general, fruits are less conductive than meat samples, and lean meat is more conductive than fat. Commercial applications exist for ohmic heating and sterilization of solid-liquid mixtures by ohmic heating have been tested.

There is a lack of proper data demonstrating the changes in major nutrients in food products and quantifying the advantages of ohmic heating. For the information according to the nutrient changes in ohmic heating, one can compare the information concerning microwave heating.

From the economical point of view, ohmic heating operational costs were found to be comparable to those for freezing and retorting of low-acid products.

High Frequency/Radio Frequency Heating

High frequency heating is done in the MHz part of the electromagnetic spectrum. The frequencies of 13.56 and 27.12 MHz are set aside for industrial heating applications. Foods are heated by transmitting electromagnetic energy through the food placed between an electrode and the ground. The main advantages of radio frequency (RF) heating compared to conventional one are improved food quality: more uniform heating, increased throughput, shorter processing lines, improved energy efficiency, improved control (heating can be controlled very precisely: switch on-switch off). Although the heating mechanism is essentially the same as with ohmic heating, RF does not require electrodes to be in contact with the food) (contactless heating). Therefore RF heating can be applied to solid as well as liquid food products. The advantage is increased power penetration. The longer wavelength at RF compared to microwave frequencies mean that RF power will penetrate further in the most products than microwave power. This can be advantage especially when thawing frozen products. There is also simpler construction than microwave systems. Improved moisture levelling corresponds with higher quality final products.

The main disadvantages are equipment and operating costs: RF heating equipment is more expensive than conventional convection, radiation, steam heating or ohmic heating systems. Nevertheless, when factors such as increased energy efficiency and increased throughput are taken into account, the total energy cost may be comparable to a conventional system. RF heating has been used in food processing industry for many decades. In particular, RF post-baking of biscuits and cereals and RF drying of foods are well-established applications. RF pasteurization and sterilization processes are becoming more important in pre-packaged food industry.

Microwave Heating

The transfer of microwave energy to food is done by contactless wave transmission. Microwaves used in the food industry for heating are frequencies 2,450 MHz or 915 MHz corresponding to 12 or 34 cm wavelength. When a microwave is applied to a food, the water molecules of the food heat up (frictional heat created by the rapidly moving dipoles in the water).

The increase in temperature of the water molecules heats surrounding components of the food by conduction and/or convection. In microwave heating less water is needed so that less extraction of valuable nutrients including minerals occur. One major limitation for industrial application of microwave heating for sterilization is the difficulty in controlling heating uniformity caused by the limited penetration depth of microwaves. The parameters important for the heating uniformity are food composition and geometry, packaging geometry and composition and applicator design (microwave energy feed system). To minimise temperature variation and also for process economy reasons, microwaves should be used in combination with conventional heating, using rapid volumetric heating for the final burst of 10-30℃ to achieve HTST-like processing.

Ultra High Hydrostatic Pressure (UHHP)

Ultra-high pressure sterilization technology is to put packaged or bulk food materials into an ultra-high pressure device. The pressure medium (water or mineral oil, etc.) is used as the pressure transmission medium, and the ultra-high static pressure of 100-1,000 MPa is applied, and the pressure is maintained at room temperature or lower temperature for a certain period of time to achieve the purpose of sterilization. The basic principle is the lethal effect of stress on microorganisms. High pressure causes many changes in the physical form, genetic mechanism, biochemical reaction and cell wall membrane of microorganisms, which affects the original physiological activities of microorganisms. Even the original function is destroyed, or irreversible changes occur. The scope of ultra-high pressure treatment only affects the non-covalent bonding in the three-dimensional structure of biopolymers. Therefore, it has no effect on the nutrition and flavor substances such as vitamins in the food, and maintains its original nutrients to the greatest extent, and is easily digested and absorbed by the human body, and it will eliminate the formation or destruction of covalent bonds caused by traditional thermal processing. Discoloration, yellowing, and unpleasant ignorance during heating, such as heat and odor. The ultra-high pressure treatment process is a pure physical process, instant compression, uniform effect, safe and hygienic operation, no industrial "three wastes", low energy consumption, and is conducive to the protection of the ecological environment and the promotion of sustainable development strategies. Therefore, as a non-thermal sterilization technology. Because of its unique and novel method, simple and easy operation, it is a technology with good application prospects.

Pulsed Electric Field

Pulse electric field (PEF) processing involves passing a high-voltage electric field (10-80 kV/cm) through a liquid material held between two electrodes in very fast pulses typically of 1-100 nanosecond duration. Microbial cells which are exposed to an external electrical field for a few microseconds respond by an electrical breakdown and local structural changes of the cell membrane. This leads to a loss of viability. Inactivation strongly depends on the intensity of the pulses in terms of field strength, energy and number of pulses applied on the microbial strain and on the properties of the food matrix. The main benefits of processing foods with short pulses of high electric fields are the very rapid inactivation of vegetative microorganisms at moderate temperatures (below 40℃ or 50℃) and with small to moderate energy requirements (50-400 J/mL). At the moment the most successful applications are for liquid foods only and there are several limitations for the use of pulsed electric as a non-thermal technology for food preservation. Some of these limitations may be solved with product formulation (less salt, less viscous, smaller particles), improved equipment design etc.

Oscillating Magnetic Fields

Static magnetic field (SMF) and oscillating magnetic field (OMF) can kill microorganisms. The strength of the stationary magnetic field does not change with time, and the strength of the magnetic field is the same in all directions. The oscillating magnetic field acts in the form of pulses, each pulse changes direction, and the magnetic field strength decays to 10% of the initial value over time.

The mechanism of magnetic field killing microorganisms is generally considered to be the strong current generated by the magnetic field. On the one hand, it interferes with the charge distribution of the cell membrane of the microorganisms, and then affects the entry and exit of the cell. On the other hand, it ionizes the substance and water in the cell to produce superoxide, thereby causing protein and enzyme denaturation, and ultimately destroy the cell structure. The magnetic flux density for killing microorganisms with an oscillating magnetic field is 5-50 T, and the pulse duration is between 10 μs and several milliseconds. The maximum frequency does not exceed 500 MHz, because the temperature rises above this value and starts to become significant.

Ultrasonic Waves

High frequency alternating electric currents can be converted into ultrasonic waves via an ultrasonic transducer. These ultrasonic waves can be amplified and applied to liquid media by an ultrasonic probe which is immersed in the liquid or an ultrasound bath filled with the treatment liquid. The antimicrobial effect of ultrasonication is due the cavitation which produces intense localized changes in pressure and temperature causing shear-induced breakdown of cell walls, disruption and thinning of cell membranes and DNA damage via free radical production. Ultrasound used alone has been stated to be insufficient to inactivate many bacterial species and would therefore not be effective as a method for food preservation alone. However, it might have in some cases synergistic effects with other methods of food preservation like heat and pressure.

Electromagnetic Radiation

Radiation is defined as the emission and propagation of energy through space or a material. From a food preservation point of view primary interest is in the electromagnetic spectrum. The electromagnetic spectrum contains different forms of radiation that differ in penetrating power, frequency, and wavelength; gamma radiation, ultraviolet radiation, infrared and microwaves are of special interest in the food industry. In this lesson the microwaves are discussed in section dealing with electric heating methods.

Irradiation

Irradiation of bulk or prepacked foods is achieved by exposing the product to a source of ionizing energy typically cobalt-60. Irradiation is not allowed in organic food processing and its use as a preservation method of conventional products is restricted in EU (mainly allowed only for spices, in Netherlands, France and Spain for frozen fish, poultry).

Ultraviolet Light/Radiation

Disinfection by ultraviolet radiation is a physical process defined by the transfer of electromagnetic energy from a light source to an organism's genetic cellular material. The lethal effects of this energy are the cell's inability to replicate. The effectiveness of the radiation is a direct function of the quantity of energy (dose) that was absorbed by the organism.

Short-wave ultraviolet light (UVC) is reported to be an effective method for inactivating bacteria on surfaces of food and on liquids like fruit and vegetable juices. Short-wave ultraviolet light has very low penetrability into solid materials. Therefore UVC treatment may be effective for disinfecting surfaces. Short-wave ultraviolet irradiation can be used to treat food surfaces. It has been used to control *Bacillus stearothermophilus* growth in thin layers of sugar. An important factor influencing the efficacy of UV treatment is the form in which the liquid makes contact with UV radiation. Because of the viscosity of most liquids containing solids (sugars, salt, starch, and other solids), the flow will be laminar, which requires a different design from a typical water unit designed to produce turbulent flow. Short-wave ultraviolet irradiation application to eliminate pathogens from fruit juices depends on ensuring that the flow of the juice is turbulent rather than laminar.

The benefits of UV in comparison to other methods of disinfection are that no chemicals are used; it is a non-heat-related process; there is no change in colour, flavour, odour, or pH; and no residuals are left in the fluid stream. It is evident that the food industry is viewing UVC-technology with special interest since there is a need to produce microbiologically safe foods while improving retention of natural flavour, colour, and appearance. One technological innovation in ascendancy is the use of UV-light for juice pasteurization; several companies are evaluating and testing UV-treatments as an alternative to pasteurizing fruit and vegetable juices, as well as other fluid products.

Infrared

Infrared (IR) waves occupy that part of the electromagnetic spectrum with frequency beyond that of visible light. In contact with material, the IR waves are either reflected, transmitted or absorbed. Absorbed waves are transformed into heat and the temperature of the material increases. The main commercial applications of IR heating are drying low moisture foods such as breadcrumbs, coca, flour, grains, malt and tea. It is also used as an initial heating stage to speed up the initial increase in surface temperature.

New Words

irreversible [ˌɪrɪ'vɜːsəbl]	*adj.*	不可逆转的；无法复原的；不能倒转的
oscillating [asə'leɪtɪŋ]	*adj.*	振荡的
photodynamic [ˌfəutəudaɪ'næmɪk]	*adj.*	光动力的；光力学的
be set aside		被搁置一边
electrode [ɪ'lektrəud]	*n.*	电极；电焊条
frictional ['frɪkʃənəl]	*adj.*	摩擦的；由摩擦而生的
dipole ['daɪpəul]	*n.*	偶极
geometry [dʒi'ɒmətri]	*n.*	几何；几何学
ultrasonication [ʌltrəsɒnɪ'keɪʃən]	*n.*	超声波处理；超声波
Bacillus stearothermophilus		嗜热脂肪芽孢杆菌
laminar ['læmɪnə]	*adj.*	薄片状的；由薄片组成的；层流的；层状的；板状的

Notes

1) The frequencies of 13.56 and 27.12 MHz are set aside for industrial heating applications.
   参考译文：13.56 MHz 和 27.12 MHz 的频率专门用于工业加热应用。
2) Irradiation of bulk or prepacked foods is achieved by exposing the product to a source of ionizing energy typically cobalt-60. Irradiation is not allowed in organic food processing and its use as a preservation method of conventional products is restricted in EU (mainly allowed only for spices, in Netherlands, France and Spain for frozen fish, poultry).
   参考译文：散装或预包装食品的辐射是通过将产品暴露于电离能源（通常为钴-60）来实现的。在欧盟，有机食品加工中不允许使用辐射，并限制其作为传统产品的保存方法（仅允许用于香料，荷兰、法国和西班牙允许用于冷冻鱼、家禽）。

Lesson 3 New Package Technology

Packaging is one of the most important processes to maintain the quality of food products for storage, transportation and end-use. It prevents quality deterioration and facilitates distribution and marketing. The basic functions of packaging are protection, containment, information and convenience. A good package can not only preserve the food quality but also significantly contribute to a business profit. Beyond the major function of preservation, packaging also has secondary functions—such as selling and sales promotion. However, the main function of food packaging is to achieve preservation and the safe delivery of food products until consumption. During distribution, the quality of the food product can deteriorate biologically and chemically as well as physically. Therefore, food packaging contributes to extending the shelf life and maintaining the quality and the safety of the food products.

Yokoyama (1985) suggested the conditions necessary to produce appropriate packaging:

a. Mass production.

b. Reasonable and efficient packaging material.

c. Suitable structure and form.

d. Convenience.

e. Consideration of disposal.

Therefore, according to these conditions, packaging design and development includes not only the industrial design fields, creativity and marketing tools, but also the areas of engineering and environmental science. Preservation, convenience and the other basic functions of packaging are certainly important, but its disposal should be treated as an important aspect of packaging development. This is a problem in package development that may confront us in the future.

The food industry uses a lot of packaging materials, and thus even a small reduction in the amount of materials used for each package would result in a significant cost reduction, and may improve solid waste problems. Packaging technology has attempted to reduce the volume and/or weight of materials in efforts to minimize resources and costs. Several trends in the food packaging evolution have been remarkable, including source reduction, design improvement for convenience and handling, and environmental concerns regarding packaging materials and processes. Food packaging has evolved from simple preservation methods to convenience, point-on-purchase (POP) marketing, material reduction, safety, tamper-proofing, and environmental issues (Table 7.1). Since the World Trade Center tragedy in 2001, food technologists have focused their attention on revising packaging systems and package designs to increase food safety and security. The level of concern regarding the use of food and water supplies as a form of bioterrorism has increased. Therefore, many applications of active packaging will be commercially developed for the security and safety enhancement of food products.

Table 7.1 Trends in the evolution of food packaging

Period	Functions and issues
1960s	Convenience, POP marketing
1970s	Lightweight, source reduction, energy saving
1980s	Safety, tamper-evidence
1990s	Environmental impact
2000s	Safety and security

Although food packaging has evolved in its various functions, every package still has to meet the basic functions. Food packaging reduces food waste and spoilage during distribution, and decreases the cost of preservation facilities. It extends the shelf life of foods, and provides safe foods to the consumers. A good package has to maintain the safety and quality of foods as well as being convenient, allowing sales promotion, and addressing environmental issues.

The quality of the packaged food is directly related to the food and packaging material attributes. Most food products deteriorate in quality due to mass transfer phenomena, such as moisture absorption, oxygen invasion, flavor loss, undesirable odor absorption, and the migration of packaging components into the food. These phenomena can occur between the food product and the atmospheric environment, between the food and the packaging materials, or among the heterogeneous ingredients in the food product itself. Therefore, mass transfer studies on the migration of package components and food ingredients, on the absorption and desorption of volatile ingredients, flavors and moisture, on gas permeation, and on the reaction kinetics of oxidation and ingredient degradation are essential for food packaging system designs.

Year after year, technology becomes better. Most developments in the field of food technology have been oriented towards processing food products more conveniently, more efficiently, at less cost, and with higher quality and safety levels. Traditional thermal processes have offered tremendous developments in the food processing industry; these include commercial sterilization, quality preservation, shelf-life extension and safety enhancement. Extended shelf-stable products manufactured by retorting or aseptic processing are available in any grocery store and do not require refrigeration. These types of products are very convenient at any place or time and are easy to handle, therefore benefiting producers, processors, distributors, retailers and consumers. The major function of extended shelf-stable food packaging is barrier protection against the invasion of microorganisms.

Beyond this simple barrier function, there has been more research and development regarding the introduction of new purposes to food packaging systems. Among these, significant new functional packaging systems include active packaging, MAP and edible films/coatings.

The development of new packaging functionalities has been possible because of technological advances in food processing, packaging material science and machinery. Among the many new technologies the development in processing and packaging machinery is notable, leading to higher standards of regulation, hygiene, health and safety. New software and part installations in unit operations have increasingly been introduced, and high-speed automation has been achieved by using new servomotors and software technologies such as the machine vision system (Tucker, 2003). The processing and packaging equipment has new functions that have increased safety, quality and productivity, and therefore it seems that the development of new packaging functions may go hand in hand with the development of new processes, materials and equipments. Packages may have new purposes if new functional packaging materials and/or materials containing functional inserts are used. Developing new packaging technologies not only implies new materials but also new packaging design systems.

Extra Active Functions of Packaging Systems

Many new "extra" functions have been introduced in active packaging technologies, including oxygen-scavenging and intelligent functions, antimicrobial activity, atmosphere control, edibility, biodegradability etc. Food packaging performs beyond its conventional protective barrier property. The new active packaging systems increase security, safety, protection, convenience, and information delivery. Active packaging systems extend the shelf life of food products by maintaining their quality longer, increase their safety by securing foods from pathogens and bioterrorism, and enhance the convenience of food processing, distribution, retailing and consumption.

There are many applications of active packaging technologies, several of which have been commercialized and are used in the food industry; these include oxygen-scavenging, carbon dioxide-absorbing, moisture-scavenging (desiccation) and antimicrobial systems. Oxygen-scavenging systems have been commercialized in the form of a sachet that removes oxygen. An oxygen-free environment can prevent food oxidation and rancidity, and the growth of aerobic bacteria and moulds. Carbon dioxide-scavenging packaging systems can prevent packages from inflating due to the carbon dioxide formed after the packaging process—for example, packaged coffee beans may produce carbon dioxide during storage as a result of non-enzymatic browning reactions. Fermented products such as kimchi (lactic acid fermented vegetables), pickles, sauces, and some dairy products can produce carbon dioxide after the packaging process. Carbon dioxide-scavenging systems are also quite useful for the products that require fermentation and aging processes after they have been packed. Moisture-scavenging systems have been used for a very long time for packaging dried foods, moisture-sensitive foods, pharmaceuticals and electronic devices; in these systems, desiccant materials are included in the package in the form of a sachet. Recently, the sachets have contained humectants as well as desiccants to control the humidity inside the package more specifically. Moisture-scavenging systems that are based on desiccation are evolving to control the moisture by maintaining a specific relative humidity inside the package by absorbing or releasing the moisture.

Antimicrobial packaging applications are directly related to food microbial safety and bioterrorism, as well as to shelf life extension by preventing the growth of spoilage and/or pathogenic microorganisms. The growth of spoilage microorganisms reduces the food shelf life, while the growth of pathogenic microorganisms endangers public health. Antimicrobial packaging systems consist of packaging materials, in-package atmospheres and packaged foods, and are able to kill or inhibit microorganisms that cause food-borne illnesses.

Intelligent packaging has been categorized both as a part of active packaging and as a separate entity, depending on different viewpoints. It contains intelligent functions which have been researched to enhance convenience for food manufacturing and distribution and, increasingly, to improve food security and safety verification.

Modified Atmosphere Packaging

MAP is traditionally used to preserve the freshness of fresh produce, meats and fish by controlling their biochemical metabolism—for example, respiration. Nitrogen flushing, vacuum packaging and carbon dioxide injection have been used commercially for many years. However, current research and development has introduced new modified atmosphere technologies such as inert gas (e.g., argon) flushing for fruits and vegetables, carbon monoxide injection for red meats, and high oxygen flushing for red meats. For a MAP system to work effectively, optimal packaging material with proper gas permeability properties must be selected. The use of MAP systems is attractive to the food industry because there is a fast-growing market for minimally processed fruits and vegetables, non-frozen chilled meats, ready-to-eat meals and semi-processed bulk foods.

MAP dramatically extends the shelf life of packaged food products, and in some cases food does not require any further treatments or any special care during distribution. However, in most cases extending shelf life and maintaining quality requires a multiple hurdle technology system—for example, introducing temperature control as well as MAP is generally essential to maintain the quality of packaged foods. Hurdle technology is therefore important for MAP applications, since the modified atmosphere provides an unnatural gas environment that can create serious microbial problems such as the growth of anaerobic bacteria and the production of microbial toxins. Therefore, an included temperature control system is very important for quality preservation and microbial control.

Edible Films and Coatings

The use of edible films and coatings is an application of active food packaging, since the edibility and biodegradability of the films are extra functions that are not present in conventional packaging systems. Edible films and coatings are useful materials produced mainly from edible biopolymers and food-grade additives. Most biopolymers are naturally existing polymers, including proteins, polysaccharides (carbohydrates and gums), and lipids. Plasticizers and other additives are included with the film-forming biopolymers in order to modify film physical properties or to create extra functionalities.

Edible films and coatings enhance the quality of food products by protecting them from physical, chemical, and biological deterioration. The application of edible films and coatings is an easy way to improve the physical strength of the food products, reduce particle clustering, and enhance the visual and tactile features of food product surfaces. They can also protect food products from oxidation, moisture absorption/desorption, microbial growth, and other chemical reactions. The most common functions of edible films and coatings are that they are barriers against oils, gas or vapours, and that they are carriers of active substances such as antioxidants, antimicrobials, colors and flavors. Thus, edible films and coatings enhance the quality of food products, which results in an extended shelf life and improved safety.

A continuing trend in food packaging technology is the study and development of new materials that possess very high barrier properties. High-barrier materials can reduce the total amount of packaging materials required, since they are made of a thin or lightweight materials with high-barrier properties. The use of high-barrier packaging materials reduces the costs in material handling, distribution/transportation and waste reduction.

Convenience is also a "hot" trend in food packaging development. Convenience at the manufacturing, distribution, transportation, sales, marketing, consumption and waste disposal levels is very important and competitive. Convenience parameters may be related to productivity, processibility, warehousing, traceability, display qualities, tamper-resistance, easy opening, and cooking preparation.

A third important trend is safety, which is related to public health and to security against bioterrorism. It is particularly important because of the increase in the consumption of ready-to-eat products, minimally processed foods and pre-cut fruits and vegetables. Food-borne illnesses and malicious alteration of foods must be eliminated from the food chain.

Another significant issue in food packaging is that it should be natural and environmentally friendly. The substitution of artificial chemical ingredients in foods and in packaging materials with natural ingredients is always attractive to consumers. Many ingredients have been substituted with natural components—for example, chemical antioxidants such as BHA, BHT and TBHQ have been replaced with tocopherol and ascorbic acid mixtures for food products. This trend will also continue in food packaging system design areas. To design environmentally friendly packaging systems that are more natural requires, for example, the partial replacement of synthetic packaging materials with biodegradable or edible materials, a consequent decrease in the use of total amount of materials, and an increase in the amount of recyclable and reusable (refillable) materials.

Food science and packaging technologies are linked to engineering developments and consumer studies. Consumers tend continuously to want new materials with new functions. New food packaging systems are therefore related to the development of food-processing technology, lifestyle changes, and political decision-making processes, as well as scientific confirmation.

New Words

confront [kən'frʌnt]	v.	面对；面临；正视；处理；对峙；对抗
tamper-proofing		防篡改
grocery ['grəʊsəri]	n.	食品杂货店；食品杂货业；食品杂货
servomotor ['sɜːvəʊˌməʊtə]	n.	伺服电动机；继动器
sachet ['sæʃeɪ]	n.	香囊；小袋
pharmaceutical [ˌfɑːmə'suːtɪkl]	adj.	制药的；n. 药物
desiccant ['desɪkənt]	adj.	去湿的；使干燥的
argon ['ɑːgɒn]	n.	［化学］氩
vapour ['veɪpə (r)]	n.	蒸汽；水汽
refillable [ˌriː'fɪləbl]	adj.	适于再装的

Notes

1) Most food products deteriorate in quality due to mass transfer phenomena, such as moisture absorption, oxygen invasion, flavor loss, undesirable odor absorption, and the migration of packaging components into the food.

   参考译文：由于传质现象，如吸湿、氧气侵入、风味损失、不良气味吸收以及包装成分迁移到食品中，大多数食品的质量会恶化。

2) An oxygen-free environment can prevent food oxidation and rancidity, and the growth of aerobic bacteria and moulds.

   参考译文：无氧环境可以防止食物氧化和酸败，以及需氧细菌和霉菌的生长。

3) The most common functions of edible films and coatings are that they are barriers against oils, gas or vapours, and that they are carriers of active substances such as antioxidants, antimicrobials, colors and flavors.

   参考译文：可食用薄膜和涂层最常见的功能是，它们是油、气体或蒸汽的屏障，并且是抗氧化剂、抗菌剂、色素和香料等活性物质的载体。

Reading Material 1　Genetically Modified Food

Genetically Modified Organisms (GMOs) are being made by inserting a gene from an external source such as viruses, bacteria, animals or plants into usually unrelated species. Biotechnology has granted us the ability to overcome insurmountable physiological barriers and to exchange genetic materials among all living organisms.

The use of recombinant DNA technology has the potential to allow the creation of an organism which is desired and designed by human. Genetically Modified Food (GMF) means any food containing or derived from a genetically engineered organism. Describing biotechnology methods is beyond the scope of this reading material, however, it is informative to only name some of the vastly used techniques in creating GM crops: Agrobacterium has been used as an intermediate organism for transferring a desirable gene into plants. This has been a successful method for modification of trees and cereal crops. Biolistic transformation is a physical method by which the genes of interest are bombarded into the plant cells and DNA-coated beads are usually used as carriers.

Another technique which facilitates the incorporation of genes into the host genome is called Electroporation. This is a suitable method for plant tissues without cell walls. DNA enters the plant cells through minute pores which are temporarily caused by electric pulses. These holes can be also created by microscopic crystals. Another recent method consists of Microinjection which is direct introduction of DNA into genome. Antisense technology is also a useful method for deactivation of specific genes such as those responsible for softening of fruits and fighting against plant viral infections.

With currently available techniques the favorite DNA are inserted to only a few numbers of the treated cells. Therefore, in order to detect whether the incorporation of the gene to the cell has taken place, the desired DNA are generally attached to marker gene before their transfer. These marker genes allow researchers to verify whether transfer of the desired DNA has properly occurred. However, after the successful gene transfer, important factors that have triggered debates over the safety of GM crops are the genotypic and phenotypic stability and permanence inheritance.

The majority of the biotech-crops available on the global market have been genetically manipulated to express one of these basic traits: resistance to insects or viruses, tolerance to certain herbicides and nutritionally enhanced quality. At present, more than 148 million hectares of farmland are under cultivation for biotech crops throughout the world. There has been a 60-fold rise in the application of Agri-biotechnology since 1996, when the first biotech-crop was commercially produced. Major producers of GM crops include USA, Argentina, Canada, and China. In the US, about 80% of maize, cotton and soya are biotech varieties. In Canada Genetically Engineered (GE) ingredients are used in more than 70% of the processed food products. The current rate of biotech-crops adoption is remarkably higher in developing versus industrialized countries (21% vs. 9%). Developing countries are rapidly accepting the technology with the hope of alleviating hunger and poverty. These countries account for 40% of the global farmlands used for GM crop cultivation. It is predicted that, by 2015, more than 200 million hectares of lands will be planted by biotech-crops in about 40 countries.

The emergence of agricultural biotechnology has created social and ethical contradictions. The widespread debate exists as to how biotechnology can be used for planting high quality and high yield crops while protecting ecosystem and human health.

While it is claimed that food biotechnology, by improvement of the plant productivity and developing nutrient-fortified staple food, is the promising solution to malnutrition and food shortage, the accumulating evidence over 20 years of GMF introduction to the market does not fully support these claims. The consumers are mainly concerned about the long-term human health effects of the biotech crops such as antibiotic resistance, allergenicity, unnatural nutritional changes and toxicity. Furthermore, Agri-biotech companies and their affiliated scientists present GM food as an environmentally friendly crop.

It is excessively stated over the media and through their dependent scientific publications that GM crops containing genes expressing herbicide tolerance and pest resistance lead to reduction of broad spectrum pesticides and herbicide use. Also, they profess that GM crops help diminishing greenhouse global emissions by reducing needs for plowing (replacement of energy-intensive by low-till agriculture). On the other hand, environment talists believe that engineering of the genetic materials could deeply transform the global ecosystem from all possible aspects. They are concerned about the long-term consequences of GM agriculture on biodiversity as it may create superweeds and superpests which can potentially disturb the balance of nature and cause serious hazards for beneficial insects. In this article, different views on agricultural biotechnology which has given rise to debates between advocated and opponents of GM crop are provided.

The information presented in this reading material was collected through extensive web searches of databases such as Regulatory Framework on Food Biosafety implemented by UNEPGEF； guidelines of European Parliament's committee on the Environment, Public Health and Food Safety；Food and Agriculture Organization of the United Nations, biosafety guidelines for crop production and food labeling and also scientific data presented by independent scientists of non-profit international organizations and many others.

Major Concerns

Much of the current debates on agricultural biotechnology have focused on the potential risks of GM crops for human health. Some of the health risks pertinent to unapproved GMFs include antibiotic resistance, allergenicity, nutritional changes and the formation of toxins. To address the possible drawbacks of biotechnology application in engineered foods, we point out some of the problems stemming out from genetic modification techniques.

GE Techniques

GE techniques have been used to transfer single gene traits such as herbicide tolerance from soil microbes into plant cells. However, recent studies in higher eukaryotic cells have shown that genes do not function independently from each other. For example, it has been discovered that human genome is not a simple collection of independent genes. Genes, instead of being constant and static, are dynamic and operate in an interactive system and intertwined with one another. Furthermore, proteins do not function separately; rather they behave in interactive network systems. Gene traits work in the cell by intercommunication and reciprocity. Hence, one gene might not determine one trait, be it herbicide tolerance, or resistance to pest. Therefore, the genetic engineering techniques seem to be imprecise and must include gene optimization steps to minimize this concern. The new understanding of genome function has changed the genetic concept which launched biotech industry a couple of decades ago.

To make a GM crop, the gene of interest is inserted into the crop's genome using a vector. This vector might contain several other elements, including viral promoters, transcription terminators, antibiotic resistance and marker genes. The genes incorporated into a genome, could reside anywhere, cause mutation in the host genome, and move or rearrange after insertion or in the next generations. Transgenic DNA might break up and reintegrate into the genome again (recombination) leading to chromosomal rearrangement in successive generations and could potentially change the transgenic crops in a way to produce proteins that are allergic or cause other health problems.

As DNA does not always fully defragment in the digestive system, human gut microflora and pathogens can take up GM materials including antibiotic resistance genes. This may cause the reduction of the effectiveness of antibiotics and therefore increasing the risk of antibiotic-resistant diseases. Some scientific advices have proposed that such markers should be replaced by non-antibiotic marker system in GMF production. In this regard, the Food Safety Unit of WHO has been assessing the safety of antibiotic resistance marker genes. However, the proponent of commercial production of GMF believe that DNA are abundant in all the foods we eat, but there has not been any evidence of the gene transfer from the food source to gut bacteria.

Seed companies argue that viruses have been engineered to be dormant in plant cells and therefore they are safe. Contrary to these claims, studies have shown that viruses, lacking the gene needed for movement, can easily gain it from neighboring genes.

Health Risks Associated with GM Food Consumption

Many scientific data indicate that animals fed by GM crops have been harmed or even died. Rats exposed to transgenic potatoes or soya had abnormal young sperm; cows, goats, buffalo, pigs and other livestock grazing on Bt-maize, GM cottonseed and certain biotech corn showed complications including early deliveries, abortions, infertility and also many died. However, this is a controversial subject as studies conducted by company producing the biotech crops did not show any negative effects of GM crops on mice. Although Agri-biotech companies do not accept the direct link between the GMFs consumption and human health problems, there are some examples given by the opponents. For example, the foodborne diseases such as soya allergies have increased over the past 10 years in USA and UK and an epidemic of Morgellons disease in the US. There are also reports on hundreds of villagers and cotton handlers who developed skin allergy in India. Recent studies have revealed that Bacillus thuringiensis corn expresses an allergenic protein which alters overall immunological reactions in the body.

The aforementioned reports performed by independent GM researchers have led to a concern about the risks of GMFs and the inherent risks associated with the genetic technology. It is therefore essential that the safety and long-term effects of GM crops should be examined before their release into the food chain by all organizations responsible to produce GMFs.

In order to give the public the option of making informed decision about the consumption of GMF, enough information on the safety tests of such product is required. Unfortunately, such data are scarce due to a number of factors. For example, it is hard to compare the nutritional contents of GM crops with their conventional counterparts because the composition of crops grown in different areas might vary depending on the growth and agronomic conditions. At the present there is no peer-reviewed publication on clinical studies of GMF effects on human health.

Current testing methods being used in biotech companies appear to be inadequate. For instance, only chemical analysis of some nutrients is reported and generally consider the GM crops equal to its conventional crops when no major differences are detected between the compound compositions in both products. Such approach is argued to guarantee that the GM crop is safe enough to be patented and commercially produced. It is strongly believed that animal trials should be used to evaluate the probable toxic effects of GMFs. Herbicide and glyphosphate resistant soybeans as well as GM cotton resistant to insects are claimed to be substantially equal to conventional soybeans or cotton. However, in these studies other than the use of inappropriate statistics, instead of comparing GM crops with the control grown at same locations, samples from different areas were measured, while it is known that environmental conditions could have major effects on the components levels. Another example is from the results of toxicological studies conducted on a variety of animals fed with glyphosate-resistant soybean (GTS) which were shown to be similar for GTS fed and control group. However, these experiments were not scientifically sound since high dietary protein concentration and very low level of GTS have hidden any real effects of GM and basically these experiments were more a commercial and not scientific studies. Also, there are some false claims on the improvement of the protein content of GM crops expressing the desired protein from an inserted gene. For example, studies on GM potato and containing soybean glycine gene did not show considerable increase in the protein content or even amino acid profile and as for GM rice the rise in protein content was due to the decline in moisture rather than the increase in protein content.

Also, there are some difficulties with assessing the allergenicity of GM crops. When the gene causing allergenicity is known, such as the gene for the alpha-amylase trypsin inhibitors, or cod proteins, it is easier to recognize whether the GMF is allergenic by using in vitro tests. Of course to test the stability of GMF products in the digestive systems, human/animal trials are required and data bank studies are effective. Since insertion of a non-allergenic gene might cause over expression of already existing minor allergen, it is difficult to specifically identify whether a new GM crop with a gene transferred from a source with unknown allergenicity is allergenic before its introduction to the food chain.

Current Debates

The genetic modification of crops has been a controversial issue since the first commercial production of GMF. The proponents of such technologies claim that bio-engineering of food is absolutely safe and it is similar to what has been happening through traditional agriculture for thousands of years. However, in selective breeding when two parental plants are crossed to obtain a desirable trait, it is likely that other unpleasant characteristics are transferred as well. Therefore, taking out undesirable traits is a slow process and requires trial and errors through several generations of plants breeding. In this context, modern biotechnology has allowed us to go beyond natural physiological reproductive barriers in a manner that gene transfer among evolutionarily divergent organisms is now possible and therefore, individual genes expressing certain traits in animals or microorganisms can be precisely incorporated to the plant genome.

GM advocates believe that conventional breeding can achieve similar results using transferred gene but only within related species and in a lengthy and imprecise process. However, GMF opponents explain that genetic engineering bears no resemblance to natural breeding as it forcibly combines genes from unrelated species together; species that were perfectly separated over billions of years of evolution. They believe that the genetic engineering is not an alternative to traditional breeding as natural crossing of plants contributes thousands of genes to the offspring through the elegant dance of life.

Agri-biotech companies claim that recombinant DNA techniques can bring advantages for consumers such as nutritional enhancement as well as improving the quality and yield of food and non-food plants such as cotton and pharmaceuticals. Most of the claims about the benefits of GMF have been proposed by the seed industry. However, independent scientists warn that the publications on the success of the GM in offering more nutritious and safe food is not based on expected scientific standards.

Drug studies funded by pharmaceutical companies are more likely to report positive result in favor of the sponsor than independently funded studies. The biased results might be achieved by the type of experiment design, selection of data and briefing the actual findings to what is expected. The same might be happening with research conducted by the seed industry. The majority of research experiments on transgenic plants are being performed by the private sector and those carried out in universities are funded by the industry. Therefore, independent scientists should urgently follow strict precautionary approach in designing experiments on GMF. GM plants have to meet the criteria of the guidelines in order to get approval for entering the market.

However, the regulatory and scientific capacities to implement such guidelines need to be built up worldwide specifically in developing countries.

Intellectual Property Rights (IPR) are one of the important factors in the current debate on GMF. The GM crops are patented by Agribusiness companies leading to monopolization of the global agricultural food and controlling distribution of the world food supply. Social activists believe that the hidden reason why biotech companies are eager to produce GM crops is because they can be privatized, unlike ordinary crops which are the natural property of all humanity. It is argued for example that to achieve this monopoly, the large Agri-biotech company, Monsanto, has taken over small seed companies in the past 10 years and has become the biggest Agri-biotech Corporation in the world. The patent right for vegetable forms of life also affect the livelihoods of family farmers as they are required to sign a contract preventing them from saving and re-planting the seeds, thus they have to pay for seeds each year.

Conclusion

Taking everything into consideration, GM crops are alive; they can migrate and spread worldwide. In this regard, clear signals should be sent to biotech companies to proceed with caution and avoid causing unintended harm to human health and the environment. It is widely believed that it is the right of consumers to demand mandatory labeling of GM food products, independent testing for safety and environmental impacts, and liability for any damage associated with GM crops. We are aware that many regulatory laws already exist for risk assessments which are performed on three levels of impacts on Agriculture (gene flow, reducing biodiversity), Food and Food safety (allergenicity, toxicity), and Environment (including non target organism); And at the same time, in recent years *Cartagena Protocol* has created laws and guidelines and has obliged countries and companies to obey them for production, handling and consumption of GM materials.

Notes

1) Some of the health risks pertinent to unapproved GMFs include antibiotic resistance, allergenicity, nutritional changes and the formation of toxins.
   参考译文：与未经批准的转基因食品相关的一些健康风险包括抗生素耐药性、过敏性、营养变化和毒素的形成。
2) It is therefore essential that the safety and long-term effects of GM crops should be examined before their release into the food chain by all organizations responsible to produce GMFs.
   参考译文：因此，所有负责生产转基因食品的组织在将转基因作物释放到食物链之前，必须对其安全性和长期影响进行检查。
3) And at the same time,in recent years *Cartagena Protocol* has created laws and

guidelines and has obliged countries and companies to obey them for production, handling and consumption of GM materials.

参考译文：同时，近年来，《卡塔赫纳议定书》制定了法律和准则，并要求各国和公司在生产、处理和消费转基因材料时遵守这些法律和准则。

Exercise

1. Translate the following sentences into English.

1）目前还没有关于转基因食品对人类健康影响的临床研究的同行审查出版物。

2）自从转基因食品首次商业化生产以来，基因改造作物一直是一个有争议的问题。

3）知识产权是当前转基因食品争论的重要因素之一。

2. Translate the following sentences into Chinese.

1）In order to give the public the option of making informed decision about the consumption of GMF, enough information on the safety tests of such product is required.

2）The GM crops are patented by Agribusiness companies leading to monopolization of the global agricultural food and controlling distribution of the world food supply.

3）The patent right for vegetable forms of life also affect the livelihoods of family farmers as they are required to sign a contract preventing them from saving and re-planting the seeds, thus they have to pay for seeds each year.

Reading Material 2 Functional Food

It is widely known that foods are sources of vitamins and minerals that support bodily functions and health (e.g., breathing, energy production, immune response). Foods are mainly consumed as cell substrates (for energy, cell differentiation and proliferation) and as the basis for chemical barriers against cell oxidation. Within many types of foods, foods for specific dietary purposes (e.g., light, diet, low-fat) and those categorized as functional foods are the main categories studied in the past 20 years.

The term functional foods were first used in Japan in the 1980s, but their definition is often misunderstood because they are regulated but not legally recognized in most countries, resulting in no statutory definition. However, recently, Granato defined functional foods as industrially processed or natural foods that when regularly consumed within a diverse diet at efficacious levels have potentially positive effects on health beyond basic nutrition. Additionally, before health claims are made for certain foods, randomized, double-blind, placebo-controlled clinical trials are necessary to establish functional efficacy. This definition narrows the widespread use of the term functional such that without a proper clinical trial and substantial experimental evidence of safety (i.e., toxicology) and functionality, no fresh, unprocessed, or processed food can be regarded as functional.

In addition to their nutritional value as a conventional food, functional foods help in the promotion of optimal health conditions and may reduce the risks of one or more non-communicable diseases, such as dyslipidemia, cancer, type Ⅱ diabetes, stroke, and cardiovascular disease. However, to be functional, a food should be validated in intervention trials to comply with the regulations in each country, for example, the Brazilian Health Regulatory Agency (ANVISA) in Brazil, the EFSA in the EU, and the FDA in the US. Together with the definition of functional foods, the main inclusion criteria for a certain functional claim for an ingredient or food on a food label are food safety, free access with no need for medical prescription (or medical advice), and evidence of health benefits when regularly consumed in a balanced diet.

Although the definition and the basic criteria are understandable and comprehensive, many researchers still have misconceptions regarding terminology associated with functional foods. Some authors declare foods or ingredients functional when they are manufactured using potentially functional substances recovered from industrial by-products or by adopting certain technological processes, whereas others provide a functional food claim based on enrichment with essential minerals. Furthermore, some authors still use the phrase "prevention of diseases" even though functional foods neither prevent nor cure any diseases, as other intrinsic and extrinsic factors (e.g., genetic factors, physical inactivity, caloric density and variety, hormones, age) play an important role in the etiology of non-communicable diseases. Similarly, many authors declare foods or ingredients functional based on *in vitro* or animal-based protocols, whereas others confuse the difference between conventional and functional foods. Thus, it is important to note that functional foods are not medications, as they do not heal/cure/prevent diseases.

Consumers today demand foods that are sustainably produced and processed, deemed safe, fresh, and natural, and have nutritional value. As with all newly designed food products, functional food development is expensive, difficult, and laborious.

Functional Foods: Development and Trends

Trends in Novel Functional Food Products

The functional food market (US, Japan, Asia Pacific, and EU) is a lucrative niche of food production and is projected to grow globally. It is anticipated to reach $304.5 billion by 2020, with an average annual growth rate of 8.5%. The most common functional food products on the market include yogurt (digestive health), cereals (cardiac health), margarines/butters (cholesterol metabolism), and energy/protein bars and drinks (hunger reduction). Below, we discuss certain innovative technologies used to develop potentially functional foods as well as technological tools that deliver bioactive ingredients/compounds.

Innovative Processing Technologies as a Tool for Development of Novel Functional Foods

The development of functional foods is essential for food companies and includes the design, optimization, and development of different formulations as well as the processing techniques that are applied to food products before they are delivered to the market. For instance, the use of thermal processing has a decisive influence on the bioavailability of nutrients and bioactive compounds present in food. Over the past two decades, innovative processing technologies (e.g., HHP, pulsed electric fields, ultrasounds, microwaves) have emerged as suitable food-processing alternatives. These sustainable technologies provide better preservation of native nutrients in fruits and vegetables, avoid microbial growth, and utilize less energy, and they can be applied to the exploitation of by-products while being eco-friendly. Hence, they have positive impacts from the functional point of view as well as in the development of new functional products. There is a significant influx of various governmental funding for research, development, and implementation of such technologies in current and new food processing.

Tools to Deliver Target Compounds

The use of bioactive compounds for the development of functional foods is conditioned, in most cases, by low solubility and reduced stability and bioavailability. To overcome these problems, the use of oral administration systems based on nanoparticles or microparticles containing bioactive compounds or essential minerals can be an effective solution. In fact, these systems can be developed in liquid forms, such as gels and pastes, and solid forms. For instance, Bonat Celli &Abbaspourrad (2018), McClements (2018), and Nikmaram et al. (2017) explained the trendiest delivery systems tailored for bioactive ingredients in food systems, including microemulsions, nanoemulsions, emulsions, solid-lipid nanoparticles, liposomes, and microgel biopolymers. The use of electrospinning for the encapsulation of bioactive compounds, both hydrophilic and hydrophobic, has been shown to be a suitable approach, as it does not involve any severe conditions of temperature, pressure, or harsh chemicals. Even though a great variety of studies was carried out to develop stable systems that increase the solubility, stability, and bioavailability of bioactive compounds, one of the biggest challenges that remains is the difficulty of scaling the laboratory results up to commercial levels. Issues often occur as the ingredients or processing operations are not commercially practical. Likewise, there is the downside that after the addition of functional compounds, the matrices that incorporate bioactive compounds may lose flavor, texture, appearance, and stability. Furthermore, the development of an adequate administration system is expensive because there must be rigorous testing on humans and animals to evaluate accurately their bioavailability and potential toxicity. For example, it is necessary to evaluate whether the administration of a bioactive-rich system allows the bioactive compound(s) to be stable and biologically active during transit through the GIT and its subsequent absorption.

Bioactive Ingredients for Functional Food Development

Probiotics, Prebiotics, and Synbiotics

By definition, probiotics are ingested living microorganisms that in adequate amounts induce health benefits in the host. To be considered probiotic, the microorganism (e.g., *Lactobacillus* and *Bifidobacterium* genera) must survive an acidic environment and exposure to bile salts found in the human body while having a good capacity for absorption in the intestine and a clear link to some health marker in clinical trials. Similarly, prebiotics are consumed so that microorganisms can selectively utilize them and provide health benefits to the host. Prebiotics have the ability to improve the survival, growth, metabolism, and beneficial health activities of probiotics in the digestive system. The most known examples from the diet are non-active food constituents that move to the colon and are selectively fermented, such as lactulose, galactooligosaccharides, fructooligosaccharides, xylooligosaccharides, and inulin as well as their hydrolysates. Finally, synbiotics are a combination of probiotics and prebiotics that provides beneficial effects to the host by efficiently improving the survival of live microbes in the digestive tract compared to either probiotics or prebiotics alone. Interestingly, the best-case scenario for synbiotics is that each constituent (pro- and prebiotic) has a beneficial additive effect. The synbiotic terminology is divided into complementary (i.e., the prebiotic component is not necessarily fermented by the probiotic strain and could theoretically support other members of the gastrointestinal microbiota without registering any ecological advantage to the probiotic strain) and synergistic (i.e., prebiotic ingredient is added to a food formulation with the specific purpose of supporting the growth of the probiotic strain).

Regarding the technological applications of probiotics, prebiotics, and synbiotics, much attention is dedicated to delivery strategies that are able to increase their chemical stability and viability during shelf life. Hence, the development of solutions to improve a product's stability throughout storage conditions is required. Some of these alternatives are the manufacture of nondairy foods, microencapsulation of probiotic microorganisms and their introduction to unfermented foods, assessment of novel sources of prebiotic ingredients, assessment of different functional effects, and engineering of edible coatings for probiotic bacteria entrapment and delivery. Regardless of technology, it is important to perform clinical trials to evaluate efficacy, correct dosage, and consequence of long-term exposure to probiotics, prebiotics, and/or synbiotics in humans. Recently, paraprobiotics, which are inactivated probiotic microorganisms that are able to provide health benefits, have drawn attention. However, the mechanisms of such actions are not completely elucidated, so this requires additional *in vitro* studies and clinical trials to understand the interactions between the physiological effects of paraprobiotics in humans.

More recently, postbiotics were defined as soluble compounds or metabolic by-products that are released by bacteria during their lives or after their lysis in the GIT. Short-chain fatty acids, enzymes, peptides, endo- and exopolysaccharides, and cell-surface proteins are included in this group. Some health benefits of postbiotics include the lowering of low-density lipoprotein cholesterol (LDL-C) and antioxidant, immunomodulatory, and antimicrobial effects, as described by Aguilar-Toalá (2018) and Foo (2019). However, the mechanisms of action are not fully elucidated. Therefore, it is important to provide optimal processing parameters that guarantee the maximum production, isolation, and purification as well as correct dosage of postbiotics.

Antioxidants

Beginning in the 1930s, antioxidants (i.e., ascorbic acid and tocopherols) were solely used as additives to prevent oxidation of oils/fats in foods with high concentrations of lipids. In the 1970s, clinical trials revealed those dietary antioxidants inhibited the oxidation processes and prevented oxidative stress in related diseases. From this moment onward, the use of antioxidants soared worldwide, not only as food additives but also as dietary supplements, because of their beneficial effects in humans. Halliwell & Gutteridge stated that "an antioxidant is a substance that, when present at a low concentration compared with that of an oxidizable substrate in the medium, inhibits oxidation of the substrate." Under this classification, phenolic compounds (e.g., flavonoids, phenolic acids, stilbenoids, coumarins, lignoids), carotenoids (e.g., carotenes and xanthophylls), terpenoids (e.g., monoterpenes, triterpenes, and sesquiterpenes), and some lipids (e.g., tocopherols and tocotrienols) are considered antioxidant agents. From the physiological perspective, there are three different mechanisms for protecting cells/organs: (a) single-electron transfer, (b) hydrogen-atom transfer, and (c) transition metal chelation. According to the place of origin, antioxidants may be endogenous (e.g., catalase, glutathione reductase, superoxide dismutase) or exogenous (tocopherols, carotenoids, phenolic compounds, and terpenoids). More details on these topics can be found elsewhere.

Multiple publications have associated higher consumption of fruits and vegetables (rich with antioxidants) with a lower risk of all-cause mortality, particularly cardiovascular mortality. Additional studies reported that a significant inverse association is commonly observed for cardiovascular mortality with higher consumption of fruits and vegetables, whereas this was not associated with risk of cancer mortality. In a recent study conducted by Du et al. (2017), authors concluded that regular consumers of fruits ($\geqslant$4 days/week) versus nonregular consumers had 27% lower all-cause mortality, 34% lower CVD mortality, 17% lower digestive tract cancer mortality, and 42% lower mortality from chronic obstructive pulmonary disease. More importantly, there was a log-linear dose-response relationship between the amounts of consumed fruits and the lower mortality index.

In a prospective and observational study with septic Spanish patients (n=319), serum antioxidant capacity (SAC) was correlated with mortality rate. Researchers found that lower SAC levels during the first week of sepsis are associated with higher lipid peroxidation, sepsis severity, and sepsis mortality, and they could be used as a prognostic biomarker. Similarly, two prospective cohort studies of middle-aged and elderly Chinese adults (n=134,358) in urban

Shanghai were carried out to assess the effects of intake of ascorbic acid and β-carotene as dietary supplements (consumed continuously at least 3 times/week for longer than two months).

Authors observed significant inverse associations among consumption of total dietary β-carotene and ascorbic acid with the risk of CVD mortality.

Initially, because data for antioxidants showed that they delayed the onset of oxidative reactions both *in vitro* and *in vivo*, sales of dietary antioxidants as antiaging agents in the 2000s gained more momentum and more marketing. A large database containing the oxygen radical absorbance capacity (ORAC) of more than 1,000 foods was released by the USDA in 2012, only to be removed from their website after stating that *in vitro* measurement of antioxidant capacity (i.e., ORAC) has no physiological relevance in humans. In addition, the USDA discouraged the use of ORAC on supplements and foods for marketing purposes. Halliwell also stated that the consumption of large doses of antioxidants (e.g., pills and tablets) generally failed to prevent human diseases, in part because they do not decrease oxidative damage *in vivo*. Additionally, an overdose of dietary antioxidants could present pro-oxidant effects, thus causing deleterious effects in humans, which is not desired. For instance, with cancer patients, Khurana et al. concluded that several randomized clinical trials showed that the consumption of antioxidants during chemotherapy decreased the effectiveness of the medical treatment. Gostner et al. reviewed the pros and cons of dietary antioxidant intake and concluded that the overall uptake of antioxidants exceeds the recommended requirements. Some of these compounds include food additives such as vitamins, colorants, flavoring agents, and preservatives that are often strong antioxidants. In addition, dietary antioxidants lower leptin, thus interfering with satiety regulation and, consequently, trigger food intake.

Another main issue in food science is finding antioxidants and chemically measuring their antioxidant effectiveness by multiple *in vitro* and *ex vivo* methods. We also need to optimize their extraction conditions to obtain more target compounds from different food and herbal sources. As assessments *in vitro* chemical assays have several disadvantages and pitfalls, cell-based methods (i.e., cytotoxic and antiproliferative effects) may be the new, reliable assays for the antioxidant activity of food-based extracts. Recently, Xiao Jianbo evoked the need to strengthen the legitimacy of studies with the chemical stability of polyphenols when these substances are tested for anticarcinogenic and antiproliferative (抗癌和抗增殖) activities in cell-based methods. The author stated that some polyphenols are not stable and potentially may produce unsafe end products without any bioactivity.

Currently, continuous demand from modern consumers is pushing the food industry to decrease or even eliminate the use of synthetic additives in manufacturing; thus, numerous research and technological applications have switched to focus on natural counterparts. Hence, bioactive compounds are extracted from herbs, foods, and industrial by-products so they can be applied as antioxidants in food production. These extracts are obtained using either water or ethanol (or their binary combination) and are then assessed to identify physicochemical and functional properties. Consequently, they can be applied in the production of different foods, such as dairy products, meat-based foods, oil-in-water emulsions, edible oil, protein concentrates from marine microalgae, bread, crackers and bars, and snacks as well as packaging materials.

Although *in vitro* chemical and cell-based methods may provide a quantitative result (e.g., high in antioxidants), clinical and epidemiological data are inconclusive at this point. Thus, more research is needed to confirm or refute the hypothesis that foods rich in antioxidants can be considered functional foods. The final message should be clear: High chemical antioxidant activity measured by different *in vitro* methods may not be translated to *in vivo* antioxidant activity, and, therefore, foods with high chemical antioxidant activity cannot be automatically considered functional.

Notes

1) Foods are mainly consumed as cell substrates (for energy, cell differentiation and proliferation) and as the basis for chemical barriers against cell oxidation.

   参考译文：食物主要作为细胞基质（用于能量、细胞分化和增殖）以及作为防止细胞氧化的化学屏障的基础。

2) In addition to their nutritional value as a conventional food, functional foods help in the promotion of optimal health conditions and may reduce the risks of one or more non-communicable diseases, such as dyslipidemia, cancer, type Ⅱ diabetes, stroke, and cardiovascular disease.

   参考译文：除了作为传统食品的营养价值外，功能性食品还有助于改进健康状况，并可能降低一种或多种非传染性疾病的风险，如血脂异常、癌症、Ⅱ型糖尿病、中风和心血管疾病。

3) The use of electrospinning for the encapsulation of bioactive compounds, both hydrophilic and hydrophobic, has been shown to be a suitable approach, as it does not involve any severe conditions of temperature, pressure, or harsh chemicals.

   参考译文：使用静电纺丝封装亲水和疏水生物活性化合物已被证明是一种合适的方法，因为它不涉及任何恶劣的温度、压力或刺激性化学物质。

Exercise

1. Translate the following sentences into Chinese.

1）The most common functional food products on the market include yogurt (digestive health), cereals (cardiac health), margarines/butters (cholesterol metabolism), and energy/protein bars and drinks (hunger reduction).

2）Over the past two decades, innovative processing technologies (e.g., HHP, pulsed electric fields, ultrasounds, microwaves) have emerged as suitable food-processing alternatives.

3）The use of bioactive compounds for the development of functional foods is conditioned, in most cases, by low solubility and reduced stability and bioavailability.

References

AGREGÁN R, ECHEGARAY N, NAWAZ A, et al, 2021. Foodomic-based approach for the control and quality improvement of dairy products[J]. Metabolites, 11(12): 818.

AL-MAZEEDI H M, REGENSTEIN J M, RIAZ M N, 2013. The issue of undeclared ingredients in Halal and Kosher food production: a focus on processing aids[J]. Comprehensive Reviews in Food Science & Food Safety,12(2): 228-233.

ARCE-CORDERO J A, MONTEIRO H F, PHILLIPS H, et al, 2021. Effects of unprotected choline chloride on microbial fermentation in a dual-flow continuous culture depend on dietary neutral detergent fiber concentration[J]. Journal of Dairy Science, 104(3): 2966-2978.

AREPALLY D, REDDY R S, GOSWAMI T K, et al, 2020. Biscuit baking: a review[J]. LWT, 131: 109726.

AXELRAD D A, GOODMAN S, WOODRUFF T J, 2009. PCB body burdens in US women of childbearing age 2001—2002: an evaluation of alternate summary metrics of NHANES data[J]. Environmental Research,109 (4): 368-378.

AYED L, M'HIR S, HAMDI M, 2020. Microbiological, biochemical, and functional aspects of fermented vegetable and fruit beverages[J]. Journal of Chemistry (1): 1-12.

BARBER T M, KABISCH S, PFEIFFER A F H, et al, 2020. The health benefits of dietary fibre[J]. Nutrients, 12(10): E3209.

BERDANIER C D, DWYER J T, HEBER D, 2013. Handbook of Nutrition and Food [M]. 3rd ed. Boca Raton: CRC Press.

BORRESEN T, 2013. How to write a good scientific paper[J]. Journal of Aquatic Food Product Technology, 22(5): 434-435.

BRAGA A, GUERREIRO C, BELO I, 2018. Generation of flavors and fragrances through biotransformation and De Novo synthesis[J]. Food & Bioprocess Technology,11(12): 2217-2228.

BREDARIOL P, DE CARVALHO R A, VANIN F M, 2020. The effect of baking conditions on protein digestibility, mineral and oxalate content of wheat breads[J]. Food Chemistry, 332: 127399.

BROWN A, 2000. Understanding food principles and preparation[M]. Belmont, CA: Wadsworth Publishing Company.

COULTATE T, 2009. Food the chemistry of its components [M]. 5th ed. Cambridge: Royal Society of Chemistry Press.

DANIEL N, NACHBAR R T, TRAN T T T, et al, 2022. Gut microbiota and fermentation-derived branched chain hydroxy acids mediate health benefits of yogurt consumption in obese mice[J]. Nature Communications, 13(1): 1-18.

DHINGRA D, MICHAEL M, RAJPUT H, et al, 2012. Dietary fibre in foods: A review[J]. Journal of Food Science &Technology, 49(3): 255-266.

FLOROS J D, NEWSOME R, FISCHER W, et al, 2010. Feeding the world: today and tomorrow: the importance of food science and technology [J]. Comprehensive Reviews in Food Science and Food Safety, 9(5): 572-599.

FULLER S, BECK E, SALMAN H, et al, 2016. New horizons for the study of dietary fiber and health: a review[J]. Plant Foods for Human Nutrition, 71(1):1-12.

GRANATO D, BARBA F J, BURSAC K D, et al, 2020. Functional foods: product development, technological trends, efficacy testing, and safety[J]. Annual Review of Food Science and Technology, 11: 93-118.

HILL C, GUARNER F, REID G, et al, 2014. The international scientific association for probiotics and prebiotics consensus statement on the scope and appropriate use of the term probiotic [J]. Nature Reviews: Gastroenterology & Hepatology, 11: 506-514.

HO K L G, SANDOVAL A, 2020. Sanitation Standard Operating Procedures (SSOPs), Food Safety Engineering [M]. New York: Springer.

HOLBAN A M, GRUMEZESCU A M, 2018. Advances in Biotechnology for Food Industry[M]. London: Elsevier.

HUANG Y, MU X, WANG J, et al, 2022. The recent development of nanozymes for food quality and safety detection[J]. Journal of Materials Chemistry B, 10(9): 1359-1368.

KING H, 2013. Food safety management [M]. New York: Springer.

KUJAWSKA M, RAULO A, MILLAR M, et al, 2022. *Bifidobacterium castoris* strains isolated from wild mice show evidence of frequent host switching and diverse carbohydrate metabolism potential[J]. ISME Communications, 2(1): 1-14.

JACKSON L S, 2009. Chemical food safety issues in the United States: past, present, and future [J]. Journal of Agricultural & Food Chemistry, 57(18):8161-8170.

LEIF S, JENS R, MOGENS A, 2010. Chemical deterioration and physical instability of food and beverages[M]. Cambridge: Woodhead Publishing.

LIU J, BI J, MCCLEMENTS D J, et al, 2020. Impacts of thermal and non-thermal processing on structure and functionality of pectin in fruit-and vegetable-based products: a review[J]. Carbohydrate Polymers, 250: 116890.

LYU X, WANG X, WANG Q, et al, 2021. Encapsulation of sea buckthorn (*Hippophae rhamnoides* L.) leaf extract via an electrohydrodynamic method[J]. Food Chemistry, 365: 130481.

MAGHARI B M, ARDEKANI A M, 2011. Genetically modified foods and social concerns [J]. Avicenna Journal of Medical Biotechnology, 3(3): 109-117.

MAJERSKA J, MICHALSKA A, FIGIEL A, 2019. A review of new directions in managing fruit and vegetable processing by-products[J]. Trends in Food Science & Technology, 88: 207-219.

MARK C, 2009. Food Industry Quality Control Systems [M]. Boca Raton: CRC Press.

MOHAMMED K, 2019. Enzymes in Food Biotechnology[M]. New York: Academic Press.

MONDAL A, DATTA A K, 2008. Bread baking-a review[J]. Journal of Food Engineering, 86:465-474.

MOSES J A, NORTON T, ALAGUSUNDARAM K, et al, 2014. Novel drying techniques for the food industry[J]. Food Engineering Reviews, 6: 43-55.

MSAGATI T, 2012. Chemistry of Food Additives and Preservatives[M]. Chichester: John Wiley & Sons Ltd.

OTTEN J J, HELLWIG J P, MEYERS L D, 2006. Dietary Reference Intakes: The essential guide to nutrient requirements[M]. Washington D C: The National Academies Press.

PAN Z, YE A, DAVE A, et al, 2022. Kinetics of heat-induced interactions among whey proteins and casein micelles in sheep skim milk and aggregation of the casein micelles [J]. Journal of Dairy Science,105(5): 3871-3882.

PAUL L H, MCSWEENEY, JOHN P M, 2022. Encyclopedia of Dairy Sciences [M]. 3rd ed. New York: Academic Press.

PRESSMAN P, CLEMENS R, HAYES W, 2017. Food additive safety: A review of toxicologic and regulatory

issues[J]. Toxicology Research and Application, 1: 1-22.
RAY R C, 2020. Microbial Biotechnology in Food and Health[M]. Manhattan: Academic Press.
RAVISHANKAR R V, 2016. Advances in Food Biotechnology[M]. Chichester: John Wiley & Sons Ltd.
ROBERTSON G L, 2010. Food packaging and shelf life [M]. Boca Raton: CRC Press.
SEMYONOV D, RAMON O, SHIMONI E, 2011. Using ultrasonic vacuum spray dryer to produce highly viable dry probiotics[J]. LWT-Food Science and Technology, 44:1844-1852.
SOLIMAN G A, 2019. Dietary fiber, atherosclerosis, and cardiovascular disease[J]. Nutrients, 11(5):1155.
TORRES S, VERÓN H, CONTRERAS L, et al, 2020. An overview of plant-autochthonous microorganisms and fermented vegetable foods[J]. Food Science and Human Wellness, 9(2): 112-123.
WEXLER P, 2014. Encyclopedia of Toxicology [M]. 3rd ed. New York: Academic Press.
WU L, ZHANG C, LONG Y, et al, 2021. Food additives: from functions to analytical methods[J/OL]. Critical Reviews in Food Science and Nutrition (2021-06-01) [2022-07-27]. http://doi.org/10.1080/10408398.2021.19298231.

Glossary

A

acetaldehyde [ˌæsɪ'tældəhaɪd] *n.* 乙醛；醋醛

agar ['eɪgɑː(r)] *n.* 琼脂

agglutinate [ə'gluːtɪneɪt] *adj.* 黏着的；胶合的

v. 使…黏着；成胶状；黏合

agrochemicals [æg'rɒkemɪklz] *n.* 农业化学试剂

aldopentose ['ældəpentəʊs] *n.* 戊醛糖

ale [eɪl] *n.* 爱尔啤酒；麦芽酒

allergen ['ælədʒən] *n.* 过敏原

allergy ['ælədʒɪ] *n.* 过敏反应；过敏症；厌恶；反感

ambient ['æmbɪənt] *adj.* 周围的；外界的；环绕的

n. 周围环境

amino acid [əˌmiːnəʊ'æsɪd] *n.* 氨基酸

aminopeptidases [æmɪnoʊ'peptaɪdeɪz] *n.* 氨基肽酶

aminotransferase [æmaɪnətrænsf'reɪz] *n.* 转氨酶

amylase ['æmɪleɪz] *n.* 淀粉酶

amyloglucosidase [eɪmɪlɒgluː'kəʊsaɪdeɪz] *n.* 淀粉葡萄糖苷酶

anaerobic [æneə'rəʊbɪk] *adj.* 厌氧的，厌气的

anneal [ə'niːl] *v.* 退火；锻炼；煅烧

anthocyanin [ænθə'saɪənɪn] *n.* 花青素

anticipate [æn'tɪsɪpeɪt] *vt.* 预期；提前使用；抢…前

vi. 预言

antioxidant [æntɪ'ɒksɪdənt] *n.* 抗氧化剂；防老化剂

apoenzyme [ə'pəʊnziːm] *n.* 脱辅基酶；主酶；酶朊；脱辅基酶蛋白

aptamer ['æptəmər] *n.* 适体；适配子

aquatic [ə'kwætɪk] *adj.* 水生的；水中的；水上的

n. 水生动物；水草

aroma [ə'rəʊmə] *n.* 香味；芳香；气味；（艺术品等的）格调；韵味

ascorbate ['æskɔːbeɪt] *n.* 抗坏血酸盐

Ascorbic Acid 抗坏血酸；维生素 C

aspartame [ə'spɑːteɪm] *n.* 阿斯巴甜

aspartate [ə'spɑrteɪt] *n.* 天（门）冬氨酸盐

Aspergillus oryzae 米曲霉

atherosclerotic [ˌæθərəʊsklə'rɒtɪk] *adj.* 动脉粥样硬化的

n. 动脉粥样硬化患者

autochthonous [ɔː'tɒkθənəs]
=autochthonal；autochthonic *adj.* 土著的；土生土长的；原地形成的

autophagy [ɔː'tɑfədʒɪ]	*n.*	（细胞的）自我吞噬（作用）

B

Bacillus stearothermophilus		嗜热脂肪芽孢杆菌
bacteria [bæk'tɪəriə]	*n.*	（复数）细菌
bacteriocin [bæktɪər'ɪəʊsɪn]	*n.*	细菌素
barley ['bɑːli]	*n.*	大麦
batch [bætʃ]	*n.*	一批
	vt.	一批生产量；分批处理
be set aside		被搁置一边
benzoic acid and sodium benzoate	*n.*	苯甲酸和苯甲酸钠
beverage ['bevərɪdʒ]	*n.*	饮料
Bifidobacterium bifidum		双歧杆菌
binary fission		（原生动物、细胞等的）二分裂；二分体
bioaccessibility	*n.*	生物有效性；生物利用度
bioactive [ˌbaɪəʊ'æktɪv]	*adj.*	生物活性的
biocatalyst [bi'ːəʊkətəlɪst]	*n.*	生物催化剂
biodiversity [ˌbaɪəʊdaɪ'vɜːsətɪ]	*n.*	生物多样性
biomarker ['baɪɒmɑːkər]	*n.*	生物标志物
biscuit sandwich		饼干三明治
blackcurrant ['blækkʌrənt]	*n.*	黑醋栗；黑加仑子
bleach [bliːtʃ]	*vt.*	使漂白；使变白
blend [blend]	*vi.*	混合；协调
	n.	混合物
bottom-fermenting yeast		下面发酵酵母
brandy ['brændi]	*n.*	白兰地酒；（夹心）糖果
	v.	以白兰地酒调制
brine [braɪn]	*n.*	卤水；盐水
	vt.	用浓盐水处理（或浸泡）
buffaloe ['bʌfələu]	*n.*	水牛；野牛
buttermilk ['bʌtəmɪlk]	*n.*	酪乳；白脱牛奶；脱脂乳
butyl hydroxyanisole	*n.*	丁基羟基苯甲醚
butylated hydroxytoluene	*n.*	丁基羟基甲苯
butyric [bjuː'tɪrɪk] acid	*n.*	丁酸

C

calcium ['kælsɪəm]	*n.*	钙
canthaxanthin [kæn'θæksænθɪn]	*n.*	角黄素；斑蝥黄
caramelized ['kærəməlaɪzd]	*vt.*	（糖）熔化而变成焦糖
carbonated drinks		碳酸饮料
carcinogenic [ˌkɑːsɪnə'dʒenɪk]	*adj.*	致癌的；致癌物的
cariogenic [kæraɪəʊd'ʒenɪk]	*adj.*	生龋齿的
carmoisine [kɑːm'wɑːsiːn]	*n.*	胭脂红
carotene ['kærətiːn]	*n.*	胡萝卜素
carotenoid [kə'rɒtənɔɪd]	*n.*	类胡萝卜素
casein ['keɪsɪɪn]	*n.*	酪蛋白；干酪素

catalase ['kætəleɪs]	*n.*	过氧化氢酶
category ['kætəgərɪ]	*n.*	种类；分类；范畴
Caucasus ['kɔːkəsəs]	*n.*	高加索（俄罗斯南部地区，位于黑海和里海之间）
caulking ['kɔːkɪŋ]	*n.*	堵塞材料
cavitation [ˌkævɪ'teɪʃən]	*n.*	气穴现象；空穴作用；成穴
cell membrane	*n.*	细胞膜
cellular ['seljələ(r)]	*adj.*	细胞的；多孔的
cerebrosides [serɪb'rəʊsaɪdz]	*n.*	脑苷脂
chamber ['tʃeɪmbə(r)]	*n.*	（身体或器官内的）室，膛；房间；会所
	adj.	室内的；私人的
Champagne [ʃæm'peɪn]	*n.*	香槟酒；香槟酒色
chloroform ['klɒrəfɔːm]	*n.*	氯仿；三氯甲烷
	vt.	用氯仿麻醉
cholesterol [kə'lestərɒl]	*n.*	胆固醇
chopping ['tʃɒpɪŋ]	*adj.*	切碎的；剁碎的
		动词 chop 的现在分词形式
chronic ['krɒnɪk]	*adj.*	长期的；慢性的；惯常的
churn [tʃəːn]	*v.*	剧烈搅动；（使）猛烈翻腾；用搅乳器搅
	n.	（制作黄油的）搅乳器；（旧时）盛奶大罐
cider ['saɪdə(r)]	*n.*	苹果酒；苹果汁
Clostridium botulinum	*n.*	肉毒杆菌
coagulant [kəʊ'ægjʊlənt]	*n.*	凝固剂；促凝剂
coagulate [kəʊ'ægjuleɪt]	*vt.*	使…凝结
	vi.	凝结
coagulum [kəʊ'ægjʊləm]	*n.*	凝结物；凝固物；凝块
coalescence [ˌkəʊə'lɛsns]	*n.*	合并；联合；接合
coccus ['kɒkəs]	*n.*	球菌；小干果
coconut ['kəʊkənʌt]	*n*	椰子；椰子肉
concentration [ˌkɒnsen'treɪʃən]	*n.*	集中；聚集；专注；浓度；密度；浓缩；浓缩物
condiment ['kɒndɪmənt]	*n.*	调味品
confection [kən'fekʃən]	*n.*	糖果，蜜饯；调制；糖膏（剂）；精制工艺品
confectionery [kən'fekʃənəri]	*n.*	甜食（糖果、巧克力等）
conjunction [kən'dʒʌŋkʃn]	*n.*	结合；同时发生；关联
consistency [kən'sɪstənsi]	*n.*	连贯性；一致性；强度；硬度；浓稠度
constituent [ˌkən'stɪtuənt]	*n.*	成分；选民
	adj.	组成的；选举的
contaminate [kən'tæmɪneɪt]	*vt.*	污染；弄脏
contamination [kənˌtæmɪ'neɪʃn]	*n.*	弄脏；污染；杂质；污染物
contemporary [kən'temprəri]	*adj.*	同时代存在的；同年龄的；当代的
	n.	同代人
contributor [kən'trɪbjətə(r)]	*n.*	贡献者；捐助者；赠送者；投稿人；原因
cross-contamination	*n.*	交叉污染
cross-sectional [krɒs'sekʃənəl]	*adj.*	分类排列的
curcumin ['kɜːkjʊmɪn]	*n.*	姜黄素
curing ['kjʊərɪŋ]	*n.*	腌制
cutting-edge technologies		尖端技术
cyanide ['saɪənaɪd]	*n.*	氰化物

v. 用氰化物处理
cyclamate ['saɪkləmeɪt] *n.* 甜蜜素
cylindrical [sə'lɪndrɪkl] *adj.* 圆柱形的；圆柱体的
cytoplasm ['saɪtəʊplæzəm] *n.* 细胞质

D

deamidation [diːæmɪ'deɪʃən] *n.* 脱酰胺（作用）
defoaming agent 消泡剂
dehydrated [ˌdiːhaɪ'dreɪtɪd] *v.* 使（食物）脱水
dehydration [ˌdiːhaɪ'dreɪʃn] *n.* 脱水
denaturation [diːˌneɪtʃə'reɪʃən] *n.* 使变性；改变本性
deodorize [diː'əʊdəraɪz] *vt.* 脱去…的臭味；防臭
depolymerize [diː'pɒlɪməˌraɪz] *v.* （使）解聚
depot ['depəʊ] *vt.* 把…存放在储藏处
derivative [dɪ'rɪvətɪv] *n..* 派生物；衍生字；导数
adj. 派生的；（贬）非独创的；庸乏的
deterioration [dɪˌtɪərɪə'reɪʃn] *n.* 变质；变坏
dextran ['dekstrən] *n.* 右旋糖酐；葡萄聚糖
diabetic [ˌdaɪə'betɪk] *adj.* 糖尿病的
n. 糖尿病患者
diacetyl [daɪə'siːtɪl] *n.* 双乙酰
digestive [daɪ'dʒestɪv] *adj.* 消化的；助消化的
n. 消化药
diglycerides [daɪ'glɪsəˌraɪdz] *n.* 甘油二酯
dipole ['daɪpəʊl] *n.* 偶极
discoloration [disˌkʌlə'reiʃn] *n.* 变色；褪色；（皮肤上的）疹斑
discretionary [dɪ'skreʃənərɪ] *adj.* 任意的；自由决定的；酌情行事的；便宜行事的
discriminant [dɪ'skrɪmɪnənt] *n.* 可资辨别的因素
disinfectant [ˌdɪsɪn'fektənt] *n.* 消毒剂；杀菌剂；消毒水
adj. 消毒的
disposition [ˌdɪspə'zɪʃn] *n.* 性情；倾向；安排；处置；控制
dissociate [dɪ'səʊʃieɪt] *vt.* 分离；解离；使脱离关系
distilled beverages 蒸馏酒
DNA ligase DNA 连接酶
document ['dɒkjumənt] *n.* 文件；文献；证件；单据；文档
vt . 记录
donor cell 供体细胞
dose [dəʊs] *n.* 剂量；药量；（药的）一服；一剂
vi. 服药
dressing ['dresɪŋ] *n.* 穿衣；加工；调味品；装饰；打扮；填料；肥料
dried sausage 风干肠
dye [daɪ] *n.* 染料

E

ecology [i'kɒlədʒi] *n.* 生态学；社会生态学
ecosystem ['iːkəʊsɪstəm] *n.* 生态系统

electrode [ɪ'lektrəʊd]	*n.*	电极；电焊条
electroporation [elektrəʊpɔː'reɪʃn]	*n.*	电穿孔
elliptical [ɪ'lɪptɪkl]	*adj.*	椭圆的
embden-meyerhof-parnas (EMP) pathway		糖酵解途径
emulsifiers [ɪ'mʌlsɪfaɪəz]	*n.*	乳化剂，黏合剂
emulsify [ɪ'mʌlsɪfaɪ]	*vt.*	使…乳化
	vi.	乳化
emulsion [ɪ'mʌlʃn]	*n.*	乳状液；乳浊液；感光乳剂
endogenous [en'dɒdʒənəs]	*adj.*	内生的；内源的
endonuclease [endə'njuːklɪeɪs]	*n.*	核酸内切酶
endopeptidase [endəʊ'peptɪdeɪs]	*n.*	肽链内切酶
endoplasmic reticulum	*n.*	内质网
enterotoxin [ˌentərəʊ'tɒksɪn]	*n.*	［基医］肠毒素
erythorbate [ˌerɪ'θɔːbeɪt]	*n.*	异抗坏血酸盐
esterase ['estəreɪs]	*n.*	酯酶；甘油松香酯
esterification [eˌsterɪfɪ'keɪʃən]	*n.*	酯化作用
ether ['iːθə]	*n.*	乙醚
excessive [ɪk'sesɪv]	*adj.*	过度的；过多的
exogenous [ek'sɒdʒənəs]	*adj.*	外生的；外因的；外成的
exopeptidase [eksəʊ'peptɪdeɪs]	*n.*	肽链端解酶；外肽酶
exopolysaccharide [eksə'pɒlɪsækærɪd]	*n.*	胞外多糖
extender [ɪks'tendə]	*n.*	增量剂；补充剂；增充剂；延长器

F

fatty acids	*n.*	脂肪酸
fingerprint ['fɪŋgəprɪnt]	*n.*	指纹
	v.	取…的指印
flavonoid ['fleɪvənɔɪd]	*n.*	黄酮类；［有化］类黄酮
flavoring ['fleɪvərɪŋ]	*n.*	调味品，调味料
	v.	给…调味（flavor 的现在分词）
fluoridate ['flʊərɪdeɪt]	*v.*	加氟；在饮水中加少量之氟（以防儿童蛀牙）
food additive	*n.*	食品添加剂
foodborne ['fuːdbɔːn]	*adj.*	食物传播的；食物传染的；食源性的
food-borne disease	*n.*	食源性疾病
foot-and-mouth disease		口蹄疫
fortified ['fɔːtɪfaɪd]	*adj.*	加强的
	n.	强化酒
frictional ['frɪkʃənəl]	*adj.*	摩擦的；由摩擦而生的
fungi ['fʌŋgaɪ]	*n.*	真菌；菌类（fungus 的复数）

G

gallate ['gæleɪt]	*n.*	没食子酸盐
gas chromatography (GC)		气相色谱法
gas chromatography-mass spectrometry (GC-MC)		气相色谱-质谱联用
gelatinization	*n.*	凝胶化

gellan gum		葛兰胶
gelling agent		胶凝剂
genera ['dʒenərə]	*n.*	种类；属 genus 的复数形式
geometric [dʒi:ə'metrɪk]	*adj.*	几何学的
geometry [dʒi'ɒmətri]	*n.*	几何；几何学
gin [dʒin]	*n.*	杜松子酒；弹棉机；轧花机；陷阱；网
	v.	用轧棉机去籽；用陷阱（网）捕捉
glazed [gleɪzd]	*adj.*	像玻璃的；目光呆滞的；光滑的 动词 glaze 的过去式和过去分词
glutamine ['glu:təmi:n]	*n.*	谷氨酰胺
glutathione [glu:tə'θaɪəʊn]	*n.*	谷胱甘肽
glycerol ['glɪsərɒl]	*n.*	甘油；丙三醇
glycine ['glaɪsːn]	*n.*	甘氨酸
glycolysis [glaɪ'kɒlɪsɪs]	*n.*	糖酵解
good manufacturing practice (GMP)		良好生产规范
Gram staining		革兰染色
grinding ['graɪndɪŋ]	*adj.*	磨的；碾的；摩擦的 动词 grind 的现在分词形式

H

ham [hæm]	*n.*	火腿；火腿肉
haze [heɪz]	*n.*	烟雾
	vi.	变糊涂；变糊涂
	vt.	戏弄
hempseed ['hempˌsi:d]	*n.*	大麻籽
hexose ['heksəʊs]	*n.*	己糖
high resolution mass spectrometry (HRMS)		高分辨率质谱
high performance liquid chromatography (HPLC)		高效液相色谱法
histamine ['hɪstəmi:n]	*n.*	组织胺
histidine carboxylyase		组氨酸羧化酶
holoenzyme [hɒləʊ'enzaɪm]	*n.*	全酶
homogenization [həʊmədʒənaɪ'zeɪʃən]	*n.*	均化；均化作用
hops [hɒps]	*n.*	啤酒花（名词 hop 的复数形式）
hormonal [hɔ:'məʊnl]	*adj.*	激素的；荷尔蒙的
hydrogenated [haɪ'drɒdʒəneɪtɪd]	*adj.*	氢化的；加氢的
hydrolase ['haɪdrəleɪs]	*n.*	水解酶
hydrolysate [haɪ'drɒlɪseɪt]	*n.*	水解液；水解产物
hydrolyze ['haɪdrəlaɪz]	*vi.*	水解
	vt.	使水解
hygiene ['haɪdʒi:n]	*n.*	卫生；卫生学，保健学；环保

I

immunosuppressive [ɪ'mju:nəʊsə'presɪv]	*n.*	免疫抑制的药

	adj.	免疫抑制力的
implementation [ˌɪmplɪmen'teɪʃn]	*n.*	履行；落实；装置
imposition [ɪmpə'zɪʃən]	*n.*	强迫接受；课税；欺骗
improper [ɪm'prɒpə (r)]	*adj.*	不合适的；错误的；不道德的
in vitro		在生物体外
indigestible [ɪndɪ'dʒestəbl]	*adj.*	难消化的；难理解的
inflammatory [ɪn'flæmətrɪ]	*adj.*	炎症性的；煽动性的；激动的
ingredient [ɪn'griːdɪənt]	*n.*	成分；原料；配料；因素
inhibitor [ɪn'hɪbɪtə(r)]	*n.*	抑制剂；抗化剂；抑制者
initiate [ɪ'nɪʃieɪt]	*vt.*	开始；创始；启蒙；介绍加入
	n.	创始人
	adj.	新加入的；启蒙的
inoculate [ɪ'nɒkjuleɪt]	*vt.*	接种；嫁接；灌输
insulation [ˌɪnsju'leɪʃən]	*n.*	绝缘；隔离，孤立
intelligent packaging		智能包装
interaction [ˌɪntər'ækʃn]	*n.*	互动交流；相互影响；相互作用
interesterification [ɪntərestərɪfɪ'keɪʃən]	*n.*	酯交换；相互酯化
intestine [ɪn'testɪn]	*n.*	肠
	adj.	国内的；内部的
intricate ['ɪntrɪkət]	*adj.*	错综复杂的
ionization [ˌaɪənaɪ'zeɪʃn]	*n.*	[化学] 离子化
irreversible [ˌɪrɪ'vɜːsəbl]	*adj.*	不可逆转的；无法复原的；不能倒转的
irreversibly [ˌɪrɪ'vɜːsəbli]	*adv.*	不可逆地
isomerase [aɪ'sɒməreɪs]	*n.*	异构酶
isomerism ['aɪsəmərɪzəm]	*n.*	异构；异构现象

J

jelly ['dʒeli]	*n.*	果冻；胶状物
	vi.	结冻；做果冻
	vt.	使结冻

K

Kefir ['kefə]	*n.*	（俄）（由牛奶发酵而成的）克非尔
ketose ['kiːtəʊs]	*n.*	酮糖
koji ['kəʊdʒɪ]	*n.*	日本酒曲；清酒曲
koumiss ['kuːmɪs]	*n.*	酸马奶；马奶酒（等于 kumiss）

L

labneh		浓缩酸奶
laccase ['lækeɪs]	*n.*	漆酶；虫漆酶
lactase ['lækteɪs]	*n.*	乳糖酶
lactitol ['læktɪtɒl]	*n.*	乳糖醇
Lactobacillus acidophilus		嗜酸乳杆菌
Lactobacillus bulgaricus		保加利亚乳杆菌
lactone ['læktəʊn]	*n.*	内酯
lactose ['læktəʊs]	*n.*	乳糖

lactulose [ˌlæktʊ'ləuz]	*n.*	半乳糖苷果糖；乳果糖
lager ['lɑ:gə(r)]	*n.*	拉格啤酒；窖藏啤酒
laminar ['læmɪnə]	*adj.*	薄片状的；由薄片组成的；层流的；层状的；板状的
Langfil		瑞典酸奶
lard [lɑ:d]	*vt.*	润色；点缀；涂加猪油
	n.	猪油
lauric ['lɒrɪc] acid	*n.*	月桂酸
laxation [læk'seɪʃən]	*n.*	松弛；松懈；轻泻；轻泻剂
lecithin ['lesɪθɪn]	*n.*	卵磷脂；蛋黄素
legislate ['ledʒɪsleɪt]	*v.*	制定法律；立法
legislation [ˌledʒɪs'leɪʃn]	*n.*	法规；法律；立法；制订法律
legume ['legju:m]	*n.*	豆科植物，豆类蔬菜
lethality [lɪ'θælɪtɪ]	*n.*	致命性；杀伤力
ligase [lɪ'geɪs]	*n.*	连接酶
linoleic [lɪ'nəʊli:k] acid	*n.*	亚油酸
linolenic [lɪnə'lenɪk] acid	*n.*	亚麻酸
linseed ['lɪnˌsi:d]	*n.*	亚麻籽
lipase ['laɪpeɪz]	*n.*	脂肪酶；脂肪分解酵素
liquid chromatography (LC)		液相色谱
lubrication [ˌlu:brɪ'keɪʃən]	*n.*	润滑；润滑作用
lyase ['laɪəs]	*n.*	裂合酶；裂解酶
lysozyme ['laɪsəzaɪm]	*n.*	溶菌酶

M

macromolecule [ˌmækrəʊ'mɒləkju:l]	*n.*	大分子；高分子；巨分子
macronutrient [ˌmækrəʊ'nju:trɪənt]	*n.*	大量营养素；常量营养元素；宏量营养素
Madeira [mə'dɪərə]	*n.*	马德拉群岛（大西洋的群岛名）；马得拉白葡萄酒
malignant [mə'lɪgnənt]	*adj.*	恶性的；恶意的；恶毒的
malnutrition [mælnjʊ'trɪʃ(ə)n]	*n.*	营养失调，营养不良
malt [mɔ:lt]	*n.*	麦芽；麦芽酒
	adj.	麦芽的
	vt.	用麦芽处理
maltitol ['mɔ:ltɪtəl]	*n.*	麦芽糖醇
mannitol [mænɒl]	*n.*	甘露醇
margarine [ˌmɑ:dʒə'ri:n]	*n.*	人造黄油；人造奶油
mass spectrometry (MS)		质谱分析
mayonnaise [ˌmeɪə'neɪz]	*n.*	蛋黄酱
Mediterranean [medɪtə'reɪniən]	*n.*	地中海
	adj.	地中海的
medium ['mi:dɪəmˌ-djəm]	*n.*	方法；媒体；媒介；中间物
	adj.	中等的；适中的
melamine ['meləmi:n]	*n.*	三聚氰胺
melting points	*n.*	熔点
membrane ['membreɪn]	*n.*	细胞膜；薄膜；膜皮
Mesopotamia [mesəupə'teimjə]	*n.*	美索不达米亚（亚洲西南部）
metabolism [mə'tæbəlɪzəm]	*n.*	代谢；新陈代谢

metabolite [me'tæbəˌlaɪt] *n.* ［生化］代谢物

metabolomics [mɪtæˌbə'lɒmɪks] *n.* 代谢组学

micelle [maɪ'sel] *n.* 胶束；胶团；胶粒

micellization [mɪselaɪ'zeɪʃən] *n.* 胶束形成；胶束化

microflora [maɪkrəʊ'flɒrə] *n.* 微生物区系；微生物群落

micronutrient [maɪkrə (ʊ) 'nju:trɪənt] *n.* 微量营养素

microscope ['maɪkrəskəʊp] *n.* 显微镜

mildew ['mɪldju:] *n.* 霉；霉病
vt. 使发霉
vi. 发霉；生霉

mineral ['mɪnərəl] *n.* 矿物；矿物质；矿泉水；无机物；苏打水
adj. 矿物的；矿质的

minimally ['mɪnəməlɪ] *adv.* 最低限度地；最低程度地

miso ['mi:səʊ] *n.* 味噌；日本豆面酱

mitochondria [maɪtəʊ'kɒndrɪə] *n.* 线粒体（mitochondrion 的复数）

modified atmosphere packaging (MAP) 气调包装

molasses [mə'læsɪz] *n.* 糖蜜；糖浆

mold [məʊld] *n.* 霉菌；模子
vt. 形成；制模；发霉
vi. 发霉；符合形状

molecule ['mɒlɪkju:l] *n.* ［化学］分子；微粒；［化学］摩尔

monitored ['mɒnɪtəd] *adj.* 监视的；监控的
动词 monitor 的过去式和过去分词

monoglycerides [mɒ'nəgli:səraɪdz] *n.* 甘油一酸酯

MSNF (milk solid non-fat) 非脂乳固体

mustard ['mʌstəd] *n.* 芥末酱；芥末黄，深黄色；芥菜

mycoprotein [maɪkəʊ'prəʊti:n] *n.* 真菌蛋白；菌蛋白

myoglobin ['maɪəˌgləʊbɪn] *n.* 肌球素；肌红蛋白

N

nanomaterial [nænəʊmæ'tiəriəl] *n.* 纳米材料

natamycin [neɪtə'maɪsɪn] *n.* 纳他霉素

neotame [niə'teɪm] *n.* 纽甜

neurodegenerative [ˌnjʊːrəʊdɪ'dʒenərətɪv] *adj.* 神经变性的

neutral ['nju:trəl] *adj.* 中立的；中性的

nisin ['naɪsɪn] *n.* 乳酸链球菌肽；尼生素

nitrite ['naɪtraɪt] *n.* 亚硝酸盐

nitrosomyoglobin [naɪtrsəma'ɪəʊgləʊbɪn] *n.* 亚硝基肌红蛋白

nomenclature [nə'menklətʃə(r)] *n.* 系统命名法

non hydrolytic 非水解的

nonaqueous ['nɒn'eɪkwɪəs] *adj.* 非水的

noncovalent bond 非共价键

non-fermentable sugar 非发酵性糖

nourishment ['nʌrɪʃm(ə)nt] *n.* 食物；营养品；滋养品；养料

nuclear magnetic resonance (NMR) 核磁共振

nucleus ['nju:kliəs] *n.* 核；核心；原子核

nutrient ['nju:trɪənt]	*n.*	营养素；营养物；滋养物
	adj.	营养的；滋养的
nutrition [njʊ'trɪʃ (n)]	*n.*	营养，营养学；营养品；养分
nutritional [nju'trɪʃənəl]	*adj.*	营养的

O

off-flavours		异常风味
oleic [əʊ'li:ɪk]	*adj.*	油的；油酸的
organic [ɔ:'gænɪk]	*adj.*	有机的；组织的；器官的；根本的
	n.	有机物
organometallic [ˌɔ:gənəʊmɪ'tælɪk]	*adj.*	有机金属的
orifice ['ɒrɪfɪs]	*n.*	孔；穴；腔
oscillating [asə'leɪtɪŋ]	*adj.*	振荡的
osmotic [ɒz'mɒtɪk]	*adj.*	渗透性；渗透的
oval ['əʊvl]	*adj.*	椭圆的；卵形的
	n.	椭圆形；卵形
oxidation [ˌɒksɪ'deɪʃən]	*n.*	氧化
oxidoreductase ['ɒksɪdəʊrɪ'dʌkteɪs]	*n.*	氧化还原酶

P

palatability ['pælətəbəlɪtɪ]	*n.*	嗜食性；适口性；风味
palm [pɑ:m]	*n*	棕榈树
palmitic [pæl'mɪtɪk]	*adj.*	来自棕榈的
partial least square discriminant analysis (PLS-DA)		偏最小二乘判别分析
pasteurization [ˌpɑ:stʃəraɪ'zeɪʃn]	*n.*	加热杀菌法，巴斯德氏杀菌法
pathogen ['pæθədʒən]	*n.*	病原体；致病菌；病原菌
pathology [pə'θɒlədʒɪ]	*n.*	病理学；病状
pectin ['pektɪn]	*n.*	果胶
pectinase [pek'ti:nəz]	*n.*	果胶酶
pepsin ['pepsɪn]	*n.*	胃蛋白酶
peptidase ['peptɪdeɪs]	*n.*	肽酶
perilla [pə'rɪlə]	*n.*	紫苏属
permeability [ˌpɜ:miə'bɪləti]	*n.*	渗透性
permeable ['pɜ:miəbl]	*adj.*	能透过的；有渗透性的
pertain [pə'teɪn]	*vi.*	属于；关于；适合
pesticide ['pestɪsaɪd]	*n.*	杀虫剂；农药
pesticide residue	*n.*	农药残留
pharmaceutical [ˌfɑ:mə'su:tɪkl]	*adj.*	制药的；配药的
	n.	药品；成药
pharmacological [ˌfɑ:məkə'lɒdʒɪkl]	*adj.*	药理学的
phenol ['fi:nɒl]	*n.*	苯酚；酚类；酚类化合物
phenolic [fɪ'nɒlɪk]	*adj.*	酚的；酚醛树脂的
	n.	酚醛树脂
phenylethylamine ['fi:nɪlˌi:θaɪlə'mi:n]	*n.*	苯（基）乙胺
p-hospholipid [ˌfɒsfə'lɪpɪd]	*n.*	磷脂

photodynamic [fəʊtəʊdaɪ'næmɪk]	*adj.*	光动力的；光力学的
p-hydroxybenzoates	*n.*	对羟基苯甲酸酯；尼泊金酯类
phytochemicals [faɪtəʊ'kemɪklz]	*n.*	植物化学物；植物素（复数）
	adj.	植物化学的
phytoestrogen [ˌfeɪtəʊ'estrədʒən]	*n.*	植物雌激素
pickle ['pɪkl]	*n.*	泡菜；咸菜；腌渍物；各式腌菜
	v.	盐腌制；醋渍
pickled ['pɪkld]	*adj.*	腌渍的
pigment ['pɪgmənt]	*n.*	色素；颜料
	v.	给…着色
plasmin ['plæzmɪn]	*n.*	血纤维蛋白溶酶；血浆酶；胞浆素
plasticity [plæ'stɪsətɪ]	*n.*	塑性，可塑性；柔软性
plastid ['plæstɪd]	*n.*	质体；色素体；成形粒
poison ['pɒɪzən]	*n.*	中毒；毒害
polished rice		精米；白米
polymerase ['pɒlɪməreɪs]	*n.*	聚合酶
polypeptide [ˌpɒlɪ'peptaɪd]	*n.*	多肽；缩多氨酸
polyphenol [ˌpɒlɪ'fiːnɒl]	*n.*	多酚
pomace ['pʌmɪs]	*n.*	果渣；油渣
potherb ['pɒtˌhɜːb]	*n.*	野菜；调味香草
poultry ['pəʊltri]	*n.*	家禽；家禽肉
pourable ['pɔːrəbl]	*adj.*	流动通畅的；可浇注的
precaution [prɪ'kɔːʃn]	*n.*	预防措施；预防；防备
predominantly [prɪ'dɒmɪnəntlɪ]	*adv.*	主要地；占优势的
prefixed ['priːfɪkst]	*n.*	前缀；（人名前的）称谓
	vt.	加…作为前缀；置于前面
prerequisite [ˌpriː'rekwəzɪt]	*n.*	先决条件；前提；必备的事物
	adj.	作为前提的；必备的
preservative [prɪ'zɜːvətɪv]	*n.*	防腐剂；预防法；预防药
	adj.	保存的；有保存力的；防腐的
primer ['praɪmə(r)]	*n.*	引物
profiling ['prəʊfaɪlɪŋ]	*n.*	资料收集；剖析研究
		动词 profile 的现在分词形式
propionic acid and sodium propionate	*n.*	丙酸和丙酸钠
propyl ['prəʊpəl]	*n.*	丙烷基
protease ['prəʊtieɪz]	*n.*	蛋白酶
proteinaceous [prəʊtiː'neɪʃəs]	*adj.*	蛋白质的
provitamin ['prəʊvaɪtəmɪn]	*n.*	维生素原；前维生素
pseudomonas [(p)sju'dɒmənəs]	*n.*	［微］假单胞菌
psychrotrophs [saɪkrəʊt'rɒfs]	*n.*	耐冷菌；嗜冷菌
pudding ['pʊdɪŋ]	*n.*	布丁
pullulanase [pʌl'jʊlænəs]	*n.*	支链淀粉酶
purification [ˌpjʊərɪfɪ'keɪʃn]	*n.*	净化；提纯；涤罪
putrefactive [pjuːtrɪ'fæktɪv]	*adj.*	腐败的；腐烂的；易腐败的
pyruvate [paɪ'ruːveɪt]	*n.*	丙酮酸盐；丙酮酸酯

Q

quantification [ˌkwɒntɪfɪ'keɪʃn]	*n.*	定量；量化
quinoline yellow		喹啉黄

R

radiation [ˌreɪdɪ'eɪʃən]	*n.*	辐射；发光；放射物
rancidity [ræn'sɪdɪtɪ]	*n.*	腐败；恶臭；腐臭气味
reconstituted [ri'kɒnstɪtju:tɪd]	*adj.*	再造的；再生的
refrigeration [rɪˌfrɪdʒə'reɪʃn]	*n.*	制冷；冷藏；制冷
reinoculation [reɪnɒkjʊ'leɪʃn]	*n.*	再接种
rennin ['renɪn]	*n.*	凝乳酶（用来制干酪和凝乳食品）
retail ['ri:teɪl]	*v.*	零售；转述
	n.	零售
	adv.	以零售方式
	adj.	零售的
retard [rɪ'tɑ:d]	*vt.*	延迟；阻止
retardant [rɪ'ta:dənt]	*n.*	阻滞剂；抑止剂
	adj.	延缓的；阻碍的
revenue ['revənju:]	*n.*	税收；国家的收入；收益
ribosome ['raɪbəsəum]	*n.*	核糖体
rod [rɒd]	*n.*	棒；惩罚；枝条
rum [rʌm]	*n.*	朗姆酒；甜酒（用甘蔗或糖蜜等酿制的一种甜酒）
	adj.	古怪的；危险的；困难的

S

saccharify [sə'kærɪfaɪ]	*vt.*	使糖化
saccharin ['sækərɪn]	*n.*	糖精
Saccharomyces cerevisiae		酿酒酵母
Saccharomyces uvarum		葡萄汁酵母
safflower ['sæflauə]	*n.*	红花；［染料］红花染料
sake ['sɑ:ki]	*n.*	日本米酒，清酒
salami [sə'lɑ:mi]	*n.*	意大利香肠
salting ['sɔ:ltɪŋ]	*n.*	盐渍
sanitation [ˌsænɪ'teɪʃn]	*n.*	卫生，公共卫生；环境卫生；卫生设备
sanitation standard operating procedure (SSOP)		卫生标准操作规范
satiety [sə'taɪətɪ]	*n.*	满足；饱足；过多
sauce [sɔ:s]	*n.*	调味汁；酱
	vt.	给…调味；使…增加趣味；对…无礼
sauerkraut ['saʊəkraʊt]	*n.*	一种德国泡菜
sausage ['sɒsɪdʒ]	*n.*	香肠；腊肠
scheduled ['ʃedju:ld]	*adj.*	预定的；已排定的
seasoning ['si:zənɪŋ]	*n.*	调料品
seedling ['si:dlɪŋ]	*n.*	幼苗
sensitivity [ˌsensə'tɪvətɪ]	*n.*	感知；觉察；灵敏度；敏感

sesame ['sesəmɪ]	*n.*	芝麻
set yoghurt		凝固酸奶
shatter ['ʃætə(r)]	*v.*	粉碎；打碎；破坏；破掉
	n.	碎片；乱七八糟的状态
sherry ['ʃeri]	*n.*	雪利酒（西班牙产的一种烈性白葡萄酒）；葡萄酒
shortening ['ʃɔːtnɪŋ]	*n.*	起酥油
shrinkage ['ʃrɪŋkɪdʒ]	*n.*	收缩；减少；损失
significant [sɪg'nɪfɪkənt]	*adj.*	重要的；显著的；有重大意义的
sirup ['sɪrəp]=syrup	*n.*	糖浆
skim [skɪm]	*vt.*	略读；撇去…的浮物
	vi.	浏览
	adj.	脱脂的
skimmed [skɪmd]	*vt.*	撇去乳脂；脱脂
slime [slaɪm]	*n.*	黏液；烂泥
sodium ['səʊdiəm] nitrite		亚硝酸钠
solvent ['sɒlvənt]	*adj.*	有溶解力的
	n.	溶剂；解决方法
sorbic acid		山梨酸
sorbic acid and potassium sorbate	*n.*	山梨酸和山梨酸钾
sorbitol ['sɔːbɪtəl]	*n.*	山梨醇
soya drink		大豆饮料
soybean ['sɔɪbiːn]	*n.*	大豆；黄豆
sparkling ['spɑːklɪŋ]	*adj.*	闪闪发光的；闪烁的；起泡沫的
spectroscopy [spek'trɒskəpɪ]	*n.*	［光］光谱学
spherical ['sferɪkl]	*adj.*	球形的；球面的
sphingolipids [ˌsfɪngəʊ'lɪpɪdz]	*n.*	鞘脂类
spices [s'paɪsɪz]	*n.*	香味料；调味料；趣味，情趣
spoilage ['spɒɪlɪdʒ]	*n.*	损坏；糟蹋；掠夺；损坏物
stabilizer ['steɪbəlaɪzə(r)]	*n.*	稳定剂
staling ['steɪlɪŋ]	*n.*	老化；停滞
starter ['stɑːtə(r)]	*n.*	起始者，开创者；起始物；发酵剂
state-of-the-art techniques		最先进的技术
stearic [stɪ'ærɪk]	*adj.*	硬脂的；硬脂酸的；似硬脂的；十八酸的
sterilization [ˌsterəlaɪ'zeɪʃn]	*n.*	灭菌；消毒
steroid ['sterɔɪd]	*n.*	类固醇；甾族化合物
stevioside ['stiːvɪəsaɪd]	*n.*	甜菊糖，甜菊苷，蛇菊苷（一种非营养性的天然甜味剂，比蔗糖甜 300 倍）
sticky ends		黏性末端
stirred yoghurt		搅拌酸奶
strained yoghurt		脱乳清酸奶
Streptococcus thermophilus		嗜热链球菌
sucralose ['suːkrələʊs]	*n.*	三氯蔗糖
suffix ['sʌfɪks]	*n.*	后缀
	vt.	添后缀
sufu ['suːfʊ]	*n.*	腐乳
sulphydryl [sʌl'faɪdrɪl]	*n.*	巯基；氢硫基
sunset yellow		日落黄

superiority [suːˌpɪərɪ'ɒrətɪ]	*n.*	优越，优势；优越性
surfactant [sɜː'fæktənt]	*n.*	表面活性剂；表面活化物质
	adj.	表面活化剂的
surimi [sjuː'rɪmaɪ]	*n.*	鱼糜
susceptible [sə'septəbl]	*adj.*	易受影响的；易感动的
sweetener ['swiːtnə(r)]	*n.*	甜味剂
synthetic [sɪn'θetɪk]	*adj.*	合成的；人造的；综合的；虚伪的
	n.	合成物；人工制品

T

tallow ['tæləʊ]	*vt.*	涂脂油于；用油脂弄脏
	n.	牛脂；兽脂；动物油脂
tapioca [ˌtæpɪ'əʊkə]	*n.*	木薯淀粉
tartrazine ['tɑːtrəziːn]	*n.*	柠檬黄
termini ['tɜːmɪnaɪ]	*n.*	目的地；界标；终点（terminus 的名词复数）
terrestrial [tə'restriəl]	*adj.*	陆地的；陆生的；地球的
	n.	地球生物
tert-butyl hydroquinone	*n.*	叔丁基对苯二酚
thaumati [θɔː'mɑːtɪ]	*n.*	非洲竹芋甜素
thermal ['θɜːml]	*adj.*	热的；热量的
	n.	上升的热气流
thickener ['θɪkənə(r)]	*n.*	增稠剂
threadlike ['θredlaɪk]	*adj.*	丝状的；细长的
titrable acidity		可滴定酸度
tocopherol [tɒ'kɒfəˌrɒl]	*n.*	维生素 E；生育酚
top-fermenting yeast		上面发酵酵母
toxicity [tɒk'sɪsətɪ]	*n.*	毒性；毒力
transcriptomics [trænsk'rɪptəʊmɪks]	*n.*	转录组学
transesterification [trænsəsterəfɪ'keɪʃən]	*n.*	酯基转移
transferase ['trænsfəreɪs]	*n.*	转移酶
transformation [ˌtrænsfə'meɪʃn]	*n.*	转化；变形；改造；转变
transglutaminase [trænzgluː'tæmɪneɪs]	*n.*	转谷氨酰胺酶
trigger ['trɪgə(r)]	*v.*	触发；引起
	n.	扳机；起因
triglyceride [traɪ'glɪsəraɪd]	*n.*	甘油三酯
triphosphate [traɪ'fɒsfeɪt]	*n.*	三磷酸盐
trisaturated glycerides	*n.*	三饱和酸甘油酯
trivial ['trɪviəl]	*adj.*	琐碎的；无价值的
trypsin ['trɪpsɪn]	*n.*	胰蛋白酶
tuber ['tjuːbə(r)]	*n.*	［植］块茎；隆起

U

ubiquitous [juː'bɪkwɪtəs]	*adj.*	普遍存在的；无所不在的
ultra-pasteurized		超巴氏杀菌的
ultrasonication [ʌltrəsɒnɪ'keɪʃən]	*n.*	超声波处理；超声波
ultrasound ['ʌltrəsaʊnd]	*n.*	超声；超音波

unabridged [ˌʌnə'brɪdʒd]	*adj.*	完整的；未经删节的；足本的
unambiguous [ˌʌnæm'bɪgjuəs]	*adj.*	不含糊的
unconventional [ˌʌnkən'venʃənl]	*adj.*	非传统的；不合惯例的；非常规的
unequivocal [ˌʌnɪ'kwɪvəkl]	*adj.*	明确的；不含糊的
unicellular [ju:nɪ'seljələ(r)]	*adj.*	单细胞的
unsaturated ['ʌn'sætʃəˌreɪtɪd]	*adj.*	不饱和的

V

vacuole ['vækjuəʊl]	*n.*	液泡
veterinary ['vetnri]	*adj.*	兽医的
	n.	兽医
virus ['vaɪrəs]	*n.*	病毒
viscous ['vɪskəs]	*adj.*	黏性的；黏的
vitamin ['vɪtəmɪn]	*n.*	维生素；维他命
vodka ['vɒdkə]	*n.*	伏特加酒
volatile ['vɒlətaɪl]	*adj.*	挥发性的；不稳定的
	n.	挥发物；有翅的动物

W

water buffalo ['bʌfələʊ]	*n.*	水牛
waxes ['wæksɪz]	*n.*	蜡；蜡状物
	vt.	给…上蜡
waxy ['wæksi]	*adj.*	蜡制的；像蜡的；质地光滑的；柔软的；苍白的
well-recognized	*adj.*	公认的
well-understood	*adj.*	确定的
whey [weɪ]	*n.*	乳清；乳浆
whisky ['wɪski]	*n.*	威士忌酒
	adj.	威士忌酒的
wholesome ['həʊlsəm]	*adj.*	健全的；有益健康的；合乎卫生的；审慎的

X

xenobiotic [ˌzenəʊbaɪ'ɒtɪk]	*n.*	异型生物质
	adj.	异型生物质的
xylitol ['zaɪlɪtɒl]	*n.*	木糖醇

Y

yeast [ji:st]	*n.*	酵母

食品专业相关信息及主要期刊网址

1) FEMS Microbiology Reviews (ISSN: 0168-6445)
影响因子：16.408　　微生物学分类下的 1 区期刊
网址：http://onlinelibrary.wiley.com/journal/10.1111/(ISSN)1574-6976

2) Comprehensive Reviews in Food Science and Food Safety (ISSN: 1541-4337)
影响因子：12.811　　食品科技分类下的 2 区期刊
网址：http://onlinelibrary.wiley.com/journal/10.1111/(ISSN)1541-4337

3) Trends in Food Science & Technology (ISSN: 0924-2244)
影响因子：12.563　　食品科技分类下的 1 区期刊
网址：http://www.sciencedirect.com/science/journal/09242244

4) Critical Reviews in Food Science and Nutrition (ISSN: 1040-8398)
影响因子：11.176　　食品科技分类下的 1 区期刊
网址：http://www.tandfonline.com/toc/bfsn20/current#.VN1-adIhmlY

5) Food Hydrocolloids (ISSN: 0268-005X)
影响因子：9.147　　食品科技分类下的 1 区期刊
网址：http://www.sciencedirect.com/science/journal/0268005X

6) Food Chemistry (ISSN: 0308-8146)
影响因子：7.514　　食品科技分类下的 1 区期刊
网址：http://ees.elsevier.com/foodchem/

7) Food Reviews International (ISSN: 8755-9129)
影响因子：6.478　　食品科技分类下的 3 区期刊
网址：http://www.tandfonline.com/toc/lfri20/current#.VN2zjtIhmlY

8) Food Research International (ISSN: 0963-9969)
影响因子：6.475　　食品科技分类下的 1 区期刊
网址：http://www.journals.elsevier.com/food-research-international/

9) Journal of Food and Drug Analysis (ISSN: 1021-9498)
影响因子：6.079　　食品科技分类下的 2 区期刊
网址：http://www.journals.elsevier.com/journal-of-food-and-drug-analysis/

10) Food and Chemical Toxicology (ISSN: 0278-6915)
影响因子：6.023　　食品科技分类下的 1 区期刊
网址：http://www.journals.elsevier.com/food-and-chemical-toxicology/

11) Innovative Food Science & Emerging Technologies (ISSN: 1466-8564)
影响因子：5.916　　食品科技分类下的 1 区期刊
网址：http://www.sciencedirect.com/science/journal/14668564

12) Molecular Nutrition and Food Research (ISSN: 1613-4125)
影响因子：5.914　　　　　　食品科技分类下的 1 区期刊
网址：http://onlinelibrary.wiley.com/journal/10.1002/(ISSN)1613-4133
13) Food Quality and Preference (ISSN: 0950-3293)
影响因子：5.565　　　　　　食品科技分类下的 1 区期刊
网址：http://www.sciencedirect.com/science/journal/09503293
14) Food Control (ISSN: 0956-7135)
影响因子：5.548　　　　　　食品科技分类下的 1 区期刊
网址：http://www.sciencedirect.com/science/journal/09567135
15) Food Microbiology (ISSN: 0740-0020)
影响因子：5.516　　　　　　食品科技分类下的 1 区期刊
网址：http://www.sciencedirect.com/science/journal/07400020
16) Environmental Microbiology (ISSN: 1462-2912)
影响因子：5.491　　　　　　微生物学分类下的 1 区期刊
网址：http://onlinelibrary.wiley.com/journal/10.1111/(ISSN)1462-2920
17) Food & Function (ISSN: 2042-6496)
影响因子：5.396　　　　　　食品科技分类下的 2 区期刊
网址：http://pubs.rsc.org/en/Journals/JournalIssues/FO?e=1#!recentarticles&all
18) Journal of Food Engineering (ISSN: 0260-8774)
影响因子：5.354　　　　　　工程:化工分类下的 1 区期刊
网址：http://www.sciencedirect.com/science/journal/02608774
19) Journal of Agricultural and Food Chemistry (ISSN: 0021-8561)
影响因子：　5.279　　　　　农业综合分类下的 1 区期刊
网址：http://pubs.acs.org/journal/jafcau
20) International Journal of Food Microbiology (ISSN: 0168-1605)
影响因子：5.277　　　　　　食品科技分类下的 1 区期刊
网址：http://www.sciencedirect.com/science/journal/01681605
21) Meat Science (ISSN: 0309-1740)
影响因子：5.209　　　　　　食品科技分类下的 1 区期刊
网址：http://www.sciencedirect.com/science/journal/03091740
22) LWT-Food Science and Technology (ISSN: 0023-6438)
影响因子：4.952　　　　　　食品科技分类下的 1 区期刊
网址：http://www.journals.elsevier.com/lwt-food-science-and-technology/
23) Applied Microbiology and Biotechnology (ISSN: 0175-7598)
影响因子：4.813　　　　　　生物工程与应用微生物分类下的 2 区期刊
网址：http://link.springer.com/journal/253
24) Applied and Environmental Microbiology (ISSN: 0099-2240)
影响因子：4.792　　　　　　生物工程与应用微生物分类下的 1 区期刊
网址：http://aem.asm.org/

25) Journal of Food Composition and Analysis (ISSN: 0889-1575)
影响因子：4.556　　　　食品科技分类下的 2 区期刊
网址：http://www.sciencedirect.com/science/journal/08891575

26) Food and Bioproducts Processing (ISSN: 0960-3085)
影响因子：4.481　　　　生物工程与应用微生物分类下的 2 区期刊
网址：http://www.journals.elsevier.com/food-and-bioproducts-processing/

27) Food and Bioprocess Technology (ISSN: 1935-5130)
影响因子：4.465　　　　食品科技分类下的 2 区期刊
网址：http://link.springer.com/journal/11947

28) Journal of Functional Foods (ISSN: 1756-4646)
影响因子：4.451　　　　食品科技分类下的 2 区期刊
网址：http://www.sciencedirect.com/science/journal/17564646

29) International Journal of Dairy Technology (ISSN: 1364-727X)
影响因子：4.374　　　　食品科技分类下的 4 区期刊
网址：http://onlinelibrary.wiley.com/journal/10.1111/(ISSN)1471-0307

30) FEMS Microbiology Ecology (ISSN: 0168-6496)
影响因子：4.194　　　　微生物学分类下的 2 区期刊
网址：http://onlinelibrary.wiley.com/journal/10.1111/(ISSN)1574-6941

31) Journal of Dairy Science (ISSN: 0022-0302)
影响因子：4.034　　　　奶制品与动物科学分类下的 1 区期刊
网址：http://www.sciencedirect.com/science/journal/00220302

32) Research in Microbiology (ISSN: 0923-2508)
影响因子：3.992　　　　微生物学分类下的 3 区期刊
网址：http://www.sciencedirect.com/science/journal/09232508

33) Plant Foods for Human Nutrition (ISSN: 0921-9668)
影响因子：3.921　　　　食品科技分类下的 2 区期刊
网址：http://link.springer.com/journal/11130

34) International Journal of Food Science and Nutrition (ISSN: 0963-7486)
影响因子：3.833　　　　食品科技分类下的 3 区期刊
网址：http://informahealthcare.com/loi/ijf

35) Journal of Applied Microbiology (ISSN: 1364-5072)
影响因子：3.772　　　　生物工程与应用微生物分类下的 2 区期刊
网址：http://onlinelibrary.wiley.com/journal/10.1111/(ISSN)1365-2672

36) International Journal of Food Science and Technology (ISSN: 0950-5423)
影响因子：3.713　　　　食品科技分类下的 2 区期刊
网址：http://onlinelibrary.wiley.com/journal/10.1111/(ISSN)1365-2621

37) Journal of the Science of Food and Agriculture (ISSN: 0022-5142)
影响因子：3.638　　　　农业综合分类下的 1 区期刊
网址：http://onlinelibrary.wiley.com/journal/10.100 2/(ISSN)1097-0010;jsessionid=72771 054

F98B11D1A565028E5274857F.f03t04

38) Journal of the Science of Food and Agriculture (ISSN: 0022-5142)
影响因子：3.638　　　　　　农业综合分类下的 1 区期刊
网址：http://onlinelibrary.wiley.com/journal/10.1002/(ISSN)1097-0010

39) Journal of Cereal Science (ISSN: 0733-5210)
影响因子：3.616　　　　　　食品科技分类下的 2 区期刊
网址：http://www.sciencedirect.com/science/journal/07335210

40) BMC Microbiology (ISSN: 1471-2180)
影响因子：3.605　　　　　　微生物学分类下的 2 区期刊
网址：http://www.biomedcentral.com/1471-2180/

41) Enzyme and Microbial Technology (ISSN: 0141-0229)
影响因子：3.493　　　　　　生物工程与应用微生物分类下的 2 区期刊
网址：http://www.sciencedirect.com/science/journal/01410229

42) Journal of Microbiology (ISSN: 1225-8873)
影响因子：3.422　　　　　　微生物学分类下的 3 区期刊
网址：http://link.springer.com/journal/12275

43) Food Analytical Methods (ISSN: 1936-9751)
影响因子：3.366　　　　　　食品科技分类下的 2 区期刊
网址：http://link.springer.com/journal/12161

44) Journal of Industrial Microbiology & Biotechnology (ISSN: 1367-5435)
影响因子：3.346　　　　　　生物工程与应用微生物分类下的 2 区期刊
网址：http://link.springer.com/journal/10295

45) World Journal of Microbiology & Biotechnology (ISSN: 0959-3993)
影响因子：3.312　　　　　　生物工程与应用微生物分类下的 2 区期刊
网址：http://link.springer.com/journal/11274

46) Journal of Texture Studies (ISSN: 0022-4901)
影响因子：3.223　　　　　　食品科技分类下的 3 区期刊
网址：http://onlinelibrary.wiley.com/journal/10.1111/(ISSN)1745-4603

47) Foodborne Pathogens and Disease (ISSN: 1535-3141)
影响因子：3.171　　　　　　食品科技分类下的 2 区期刊
网址：http://online.liebertpub.com/FPD

48) Journal of Food Science (ISSN: 0022-1147)
影响因子：3.167　　　　　　食品科技分类下的 2 区期刊
网址：http://onlinelibrary.wiley.com/journal/10.1111/(ISSN)1750-3841

49) Chemical Senses (ISSN: 0379-864X)
影响因子：3.160　　　　　　行为科学分类下的 3 区期刊
网址：http://chemse.oxfordjournals.org/

50) Food Biophysics (ISSN: 1557-1858)
影响因子：3.114　　　　　　食品科技分类下的 3 区期刊

网址：http://link.springer.com/journal/11483

51) International Dairy Journal (ISSN: 0958-6946)
影响因子：3.032 食品科技分类下的 2 区期刊
网址：http://www.sciencedirect.com/science/journal/09586946

52) European Food Research and Technology (ISSN: 1438-2377)
影响因子：2.998 食品科技分类下的 2 区期刊
网址：http://link.springer.com/journal/217

53) Journal of Sensory Studies (ISSN: 0887-8250)
影响因子：2.991 食品科技分类下的 3 区期刊
网址：http://onlinelibrary.wiley.com/journal/10.1111/(ISSN)1745-459X

54) Journal of Bioscience and Bioengineering (ISSN: 1389-1723)
影响因子：2.894 生物工程与应用微生物分类下的 3 区期刊
网址：http://www.sciencedirect.com/science/journal/13891723

55) Letters in Applied Microbiology (ISSN: 0266-8254)
影响因子：2.858 生物工程与应用微生物分类下的 3 区期刊
网址：http://onlinelibrary.wiley.com/journal/10.1111/(ISSN)1472-765X

56) FEMS Yeast Research (ISSN: 1567-1356)
影响因子：2.796 生物工程与应用微生物分类下的 2 区期刊
网址：http://onlinelibrary.wiley.com/journal/10.1111/(ISSN)1567-1364

57) Journal of Medicinal Food (ISSN: 1096-620X)
影响因子：2.786 药物化学分类下的 3 区期刊
网址：http://online.liebertpub.com/jmf

58) Microbiology (ISSN: 1350-0872)
影响因子：2.777 微生物学分类下的 4 区期刊
网址：http://mic.sgmjournals.org/

59) International Journal of Systematic and Evolutionary Microbiology (ISSN: 1466-5026)
影响因子：2.747 微生物学分类下的 3 区期刊
网址：http://ijs.sgmjournals.org/

60) FEMS Microbiology Letters (ISSN: 0378-1097)
影响因子：2.742 微生物学分类下的 3 区期刊
网址：http://onlinelibrary.wiley.com/journal/10.1111/(ISSN)1574-6968

61) Starch-Stärke (ISSN: 0038-9056)
影响因子：2.741 食品科技分类下的 2 区期刊
网址：http://onlinelibrary.wiley.com/journal/10.1002/(ISSN)1521-379X

62) International Journal of Food Properties (ISSN: 1094-2912)
影响因子：2.727 食品科技分类下的 3 区期刊
网址：http://www.tandfonline.com/toc/ljfp20/current#.VN1x8tIhmlY

63) Journal of Food Biochemistry (ISSN: 0145-8884)
影响因子：2.720 生化与分子生物学分类下的 4 区期刊

网址：http://onlinelibrary.wiley.com/journal/10.1111/(ISSN)1745-4514

64) Journal of Food Science and Technology (ISSN: 0022-1155)
影响因子：2.701　　　　　　　食品科技分类下的 2 区期刊
网址：http://www.springer.com/food+science/journal/13197

65) Australian Journal of Grape and Wine Research (ISSN: 1322-7130)
影响因子：2.688　　　　　　　食品科技分类下的 2 区期刊
网址：http://onlinelibrary.wiley.com/journal/10.1111/(ISSN)1755-0238

66) European Journal of Lipid Science and Technology (ISSN: 1438-7697)
影响因子：2.679　　　　　　　食品科技分类下的 3 区期刊
网址：http://onlinelibrary.wiley.com/journal/10.1002/(ISSN)1438-9312

67) Flavour and Fragrance Journal (ISSN: 0882-5734)
影响因子：2.576　　　　　　　食品科技分类下的 3 区期刊
网址：http://onlinelibrary.wiley.com/journal/10.1002/(ISSN)1099-1026

68) British Food Journal (ISSN: 0007-070X)
影响因子：2.518　　　　　　　食品科技分类下的 3 区期刊
网址：http://emeraldinsight.com/loi/bfj

69) Journal of Food Quality (ISSN: 0146-9428)
影响因子：2.450　　　　　　　食品科技分类下的 3 区期刊
网址：http://onlinelibrary.wiley.com/journal/10.1111/(ISSN)1745-4557

70) Canadian Journal of Microbiology (ISSN: 0008-4166)
影响因子：2.419　　　　　　　生物工程与应用微生物分类下的 4 区期刊
网址：http://www.nrcresearchpress.com/journal/cjm

71) Food Science and Biotechnology (ISSN: 1226-7708)
影响因子：2.391　　　　　　　食品科技分类下的 4 区期刊
网址：http://www.fsnb.or.kr/

72) Journal of Microbiological Methods (ISSN: 0167-7012)
影响因子：2.363　　　　　　　生化研究方法分类下的 3 区期刊
网址：http://www.sciencedirect.com/science/journal/01677012

73) Journal of Food Process Engineering (ISSN: 0145-8876)
影响因子：2.356　　　　　　　工程化工分类下的 3 区期刊
网址：http://onlinelibrary.wiley.com/journal/10.1111/(ISSN)1745-4530

74) Journal of Microbiology and Biotechnology (ISSN: 1017-7825)
影响因子：2.351　　　　　　　生物工程与应用微生物分类下的 3 区期刊
网址：http://www.jmb.or.kr/

75) Journal of Basic Microbiology (ISSN: 0233-111X)
影响因子：2.281　　　　　　　微生物学分类下的 4 区期刊
网址：http://onlinelibrary.wiley.com/journal/10.1002/(ISSN)1521-4028

76) CYTA-Journal of Food (ISSN: 1947-6337)
影响因子：2.255　　　　　　　食品科技分类下的 3 区期刊

网址：http://www.tandfonline.com/toc/tcyt20/current#.VN1_6NIhmlY

77) American Journal of Enology and Viticulture (ISSN: 0002-9254)

影响因子：2.253　　　　　　生物工程与应用微生物分类下的 3 区期刊

网址：http://www.ajevonline.org/

78) Journal of Food Processing and Preservation (ISSN: 0145-8892)

影响因子：2.190　　　　　　食品科技分类下的 3 区期刊

网址：http://onlinelibrary.wiley.com/journal/10.1111/(ISSN)1745-4549

79) Current Microbiology (ISSN: 0343-8651)

影响因子：2.188　　　　　　微生物学分类下的 4 区期刊

网址：http://link.springer.com/journal/284

80) Annals of Microbiology (ISSN: 1590-4261)

影响因子：2.112　　　　　　生物工程与应用微生物分类下的 4 区期刊

网址：http://link.springer.com/journal/13213

81) Journal of Food Protection (ISSN: 0362-028X)

影响因子：2.077　　　　　　生物工程与应用微生物分类下的 4 区期刊

网址：http://www.foodprotection.org/publications/journal-of-food-protection/index.php

82) Food and Nutrition Bulletin (ISSN: 0379-5721)

影响因子：2.069　　　　　　食品科技分类下的 4 区期刊

网址：http://www.foodandnutritionbulletin.org/fnbhome.php

83) Bioscience, Biotechnology and Biochemistry (ISSN: 0916-8451)

影响因子：2.043　　　　　　食品科技分类下的 4 区期刊

网址：http://www.tandfonline.com/toc/tbbb20/current#.VN23QtIhmlY

84) Food Science and Technology International (ISSN: 1082-0132)

影响因子：2.023　　　　　　食品科技分类下的 3 区期刊

网址：http://fst.sagepub.com/

85) Cereal Chemistry (ISSN: 0009-0352)

影响因子：1.984　　　　　　食品科技分类下的 3 区期刊

网址：http://cerealchemistry.aaccnet.org/journal/cchem

86) Journal of Essential Oil Research (ISSN: 1041-2905)

影响因子：1.963　　　　　　食品科技分类下的 3 区期刊

网址：http://www.tandfonline.com/toc/tjeo20/current#.VN2-k9IhmlY

87) Journal of Food Safety (ISSN: 0149-6085)

影响因子：1.953　　　　　　生物工程与应用微生物分类下的 4 区期刊

网址：http://onlinelibrary.wiley.com/journal/10.1111/(ISSN)1745-4565

88) Journal of AOAC International (ISSN: 1060-3271)

影响因子：1.913　　　　　　分析化学分类下的 4 区期刊

网址：http://www.aoac.org/imis15_prod/AOAC/Publications/Journal_Of_AOAC/AOAC_Member/Publications/Journal_of_AOAC/The_Journal_of_AOAC.aspx

89) Journal of Dairy Research (ISSN: 0022-0299)
影响因子：1.904　　奶制品与动物科学分类下的2区期刊
网址：http://journals.cambridge.org/action/displayJournal?jid=DAR

90) Journal of the American Oil Chemists' Society (ISSN: 0003-021X)
影响因子：1.849　　食品科技分类下的2区期刊
网址：http://link.springer.com/journal/11746

91) Journal of the Institute of Brewing (ISSN: 0046-9750)
影响因子：1.759　　食品科技分类下的4区期刊
网址：http://onlinelibrary.wiley.com/journal/10.1002/(ISSN)2050-0416/

92) South African Journal of Enology + Viticulture (ISSN: 0253-939X)
影响因子：1.733　　食品科技分类下的4区期刊
网址：http://www.sawislibrary.co.za/dbtw-wpd/textbase/sajev.htm

93) Grasas y Aceites (ISSN: 0017-3495)
影响因子：1.650　　食品科技分类下的4区期刊
网址：http://grasasyaceites.revistas.csic.es/index.php/grasasyaceites

94) Agricultural and Food Science (ISSN: 1459-6067)
影响因子：1.375　　农业综合分类下的3区期刊
网址：http://ojs.tsv.fi/index.php/AFS

95) Czech Journal of Food Sciences (ISSN: 1212-1800)
影响因子：1.279　　食品科技分类下的4区期刊
网址：http://www.agriculturejournals.cz/web/cjfs.htm

96) Irish Journal of Agricultural and Food Research (ISSN: 0791-6833)
影响因子：1.125　　农业综合分类下的4区期刊
网址：http://www.teagasc.ie/research/journal/

97) Natural Product Communications (ISSN: 1934-578X)
影响因子：0.986　　药物化学分类下的4区期刊
网址：http://www.naturalproduct.us/

98) Italian Journal of Food Science (ISSN: 1120-1770)
影响因子：0.875　　食品科技分类下的4区期刊
网址：http://www.chiriottieditori.it/it/italian-journal-of-food-science/625

99) Food Science and Technology Research (ISSN: 1344-6606)
影响因子：0.668　　食品科技分类下的4区期刊
网址：http://www.karger.com/Journal/Home/227093

100) Acta Alimentaria (ISSN: 0139-3006)
影响因子：0.650　　食品科技分类下的4区期刊
网址：http://www.editorialmanager.com/aalim/

101) Food and Drug Law Journal (ISSN: 1064-590X)
影响因子：0.619　　食品科技分类下的4区期刊
网址：http://www.fdli.org/

102) Cereal Foods World (ISSN: 0146-6283)

影响因子：0.538　　　　　　食品科技分类下的 4 区期刊

网址：http://www.aaccnet.org/publications/plexus/cfw/pages/default.aspx

103) Food Hygiene and Safety Science (ISSN: 0015-6426)

影响因子：0.464　　　　　　食品科技分类下的 4 区期刊

网址：https://www.jstage.jst.go.jp/browse/shokueishi/

104) Milchwissenschaft - Milk Science International (ISSN: 0026-3788)

影响因子：0.416　　　　　　食品科技分类下的 4 区期刊

网址：http://www.speciation.net/Database/Journals/Milchwissenschaft--Milk-Science -International-;i404

105) Dairy Science & Technology (ISSN: 1958-5586)

影响因子：0.000　　　　　　食品科技分类下的 3 区期刊

网址：http://link.springer.com/journal/13594

106) Journal International des Sciences de la Vigne et du Vin (ISSN: 1151-0285)

影响因子：0.000　　　　　　食品科技分类下的 4 区期刊

网址：http://www.jisvv.com/

107) Agro Food Industry Hi Tech (ISSN: 1722-6996)

影响因子：0.000　　　　　　生物工程与应用微生物分类下的 4 区期刊

网址：http://www.teknoscienze.com/pages/af-journal-home.aspx#.VN11XNIhmlY

注：影响因子引自 2021 年；期刊分区引自《2021 年中国科学院文献情报中心 SCI 期刊分区表》。

主要数据库

EBSCO 数据库（EBSCO host）
网址：http://www.ebscohost.com
Elseviser 全文电子期刊数据库
网址：http://www.elsevier.com/
Wiley 全文电子期刊库（Wiley Online Library）
网址：http://onlinelibrary.wiley.com/
Springerlink
网址：http://link.springer.com/
美国生物医学信息检索系统 PubMed
网址：http://www.ncbi.nlm.nih.gov/pubmed/
美国工程索引（Engineering Village）
网址：http://www.engineeringvillage.com/
AGRIS 数据库（AGRICultural OnLineAccess）
网址：http://ovidsp.tx.ovid.com/spb/ovidweb.cgi
DOAJ 开放存取期刊集（DOAJ Open Access）
网址：http://www.doaj.org/
BMJ 科技期刊库（BMJ）
网址：http://group.bmj.com/products/journals
前沿（Frontiers Research Foundation）
网址：http://www.frontiersin.org
SAGE 科技期刊库（SAGE Journals）
网址：http://online.sagepub.com/
生物医药中心（BioMed Central）
网址：http://www.biomedcentral.com/
美国化学学会全文电子期刊（American Chemical Socity）
网址：http://pubs.acs.org
英国皇家化学学会电子期刊（Royal Society of Chemistry）
网址：http://www.rsc.org
斯坦福大学 Highwire Press
网址：http://home.highwire.org/
WordSciNet 全文期刊数据库
http://www.worldscientific.com/page/worldscinet
EPO 数据库
网址：http://worldwide.espacenet.com/

欧洲专利数据库
网址：http://ep.espacenet.com/
美国专利数据库
网址：http://patft.uspto.gov/
英国专利数据库
网址：https://www.gov.uk/government/organisations/intellectual-property-office
加拿大专利数据库
网址：http://www.ic.gc.ca/eic/site/cipointernet-internetopic.nsf/fra/accueil

主要学术机构和大学

1．主要学术机构

1）美国食品科学技术学会（Institute of Food Technologists）（http://www. ift.org/）

美国食品科学技术学会（IFT）成立于 1939 年，总部设在美国的芝加哥，是非营利的科学研究机构，1 万余名会员致力于食品科学、食品技术和工业、科学、政府的相关工作。IFT 发布各种食品工业的资料，包括食品科技和食品科学杂志，并致力于解决食品科学和技术方面的科学观点问题。

来自全球 90 多个国家的成员通过由学会建立的动态全球性研讨会平台（如全球性的科技年会、出版刊物及其他资源等）分享、学习、交流食品科学相关领域的最新知识及技术。IFT 同时就食品种植栽培、加工处理、制造加工、市场营销及全球范围内的食品安全等问题召开世界规模的年会，每年都有成千上万的食品科学家、研发专业人士、供应商、市场营销专家以及其他来自全球各地的业界人士参加，覆盖了新型健康益处、安全和产品创新、消费者最新偏好及趋势等众多话题。与 IFT 年会同期举办的食品科技博览会也是美洲地区规模最大的、最具影响力的食品添加剂、食品配料及科技方面的专业行业展会，展会集合了食品行业全球科技成果转化为产品的最新情况，反映了食品工业发展的方向和动态，代表了世界食品科技工业发展趋势。

2）国际微生物生态学会（International Society for Microbial Ecology）（http://www.isme-microbes.org/）

国际微生物生态学会（ISME）成立于 1998 年，是一个关于微生物生态学及其相关理论的非营利性科学组织。通过组织国际性的研讨会、工作小组、出版期刊等使全世界相关科技工作者交流科学信息。每两年举办一次的 ISME 研讨会是国际上最大的关于微生物生态学的盛会，主要涉及微生物生态及其多样性、微生物-微生物及微生物-宿主相互关系、进化基因组学、整合基因组及后基因组方法、微生物工程等内容。其出版学术期刊 *The ISME Journal* 是国际微生物生态学领域的顶级学术期刊，2021 年影响因子为 10.302。

3）国际食品保护协会（International Association for Food Protection）（http://www.foodprotection.org/）

国际食品保护协会是食品安全专业人士的非营利协会，简称 IAFP。1911 年 10 月 16 日，35 名来自澳大利亚、加拿大、美国的专家在美国威斯康星州的密尔沃基成立了第一届国际乳品监测协会。25 年后，该协会改名为国际乳品公共卫生协会。1947 年，更名为国际乳品与食品公共卫生协会。1966 年，更名为国际乳品、食品、环境卫生协会。1999 年 10 月，更名为国际食品保护协会。

其总部设立在美国艾奥瓦州，拥有 3 000 余名来自 50 个国家的会员，其使命是为全球食品安全专业人员提供一个可交流食品保护相关信息的平台。长期以来，国际食品保护协会一直颁发奖项用于嘉奖对加强食品安全作出突出贡献的食品企业、组织和个人。

4）国际食品科学技术联合会（International Union of Food Science and Technology）（http://www.iufost.org/）

国际食品科学技术联合会是世界食品科技工作者的国际学术团体，简称 IUFoST。其前身是食品科学小组，成立于 1960—1962 年，由澳大利亚、美国、英国、加拿大、瑞典、印度等国的食品科学家和食品工艺学家组成。中国科学技术学会于 1984 年参加了该联合会。

该联合会的宗旨是：在成员组织的科学家和专家之间进行国际合作和交流科学技术情报；支持食品科学在理论和应用领域的国际进展；提高食品加工、制造、保藏和流通技术；促进食品科学技术教育和培训。除常单独或联合有关国际学术组织主持召开主题国际学术会议以外，每 4 年支持召开一次世界食品科学技术会议，同时召开全体代表会议，进行执行委员会的换届。

该联合会是一个非营利性组织，拥有来自 65 个成员国家的 200 000 余名食品科学技术领域相关专家，每个成员国都代表了其本国的食品科学组织，为全球性的食品科学技术交流提供了平台。

5）国际谷类科学与技术协会（International Association for Cereal Science and Technology）（http://www.icc.or.at）

国际谷类科学与技术协会（ICC）于 1955 年在德国汉堡成立。它是一个由世界各地小麦及其他谷类食品研磨、面包制作及其他谷类食品制作相关领域的专家组成的国际化组织。在最近的几年，ICC 将其重点延伸到如何提高食品质量、食品安全方面。同时，为全球的谷类专家和技术者提供了一个非政治性、非营利性的研讨交流平台，在进行谷物领域信息交流的同时，促进了地区间、甚至全球的合作，同时对国际研究项目的合作和进行起到非常重要的作用。

2. 主要大学

1）普渡大学（Purdue University）（http://www.purdue.edu/）

普渡大学是一所历史悠久的研究性州立大学，世界著名高等学府。它拥有 6 个校区，学生人数约 4 万人，主校区位于美国中西部印第安纳州西拉法叶。普渡大学在国际上享有很高的声誉，尤其是理工科在世界上知名度极高。拥有 20 多名美国工程院院士，造就过 13 位理学领域的诺贝尔奖得主，是当之无愧的世界级名校。普渡大学的工程学院是全球最顶尖的工程学院之一，其在农学、计算机科学、航空航天学等领域也处于学术界领先地位。农业及生物工程连续 3 年在美国大学排行榜名列榜首。

2）康奈尔大学（Cornell University）（http://www.cornell.edu/）

康奈尔大学主校区位于美国伊萨卡小城，是一座综合研究型大学，在世界范围内享有极高的学术声誉，共有 40 多位师生曾荣获诺贝尔奖。其校训的宗旨是创建学科齐全、包罗万象的新型综合性大学，因而该校一直被誉为美国历史上第一所具有真正意义的全民大学。该校是美国常春藤盟校中唯一开设食品专业的大学，综合实力强，研究方向全面，涉及所有食品科学的研究方向。

3）加利福尼亚大学戴维斯分校（University of California, Davis）（http://ucdavis.edu/）

加利福尼亚大学戴维斯分校是美国的一所著名的公立大学，位于美国加州的戴维斯市。加利福尼亚大学戴维斯分校成立于 1908 年，是加州大学的十所分校之一。加利福尼亚大学戴维斯校区设有护理学院、农业及环境科学学院、生物科学学院、工程学院、文理学院、管理研究生院、教育学院、法学院、医学院和兽医学院。加利福尼亚大学戴维斯校区开设有农业科学、环境科学、动物学、数学、艺术学、大气科学、生物学、化学、物理学、计算机科学、工程学、传播学、经济学、教育学、语言学、历史学、法律学、音乐、哲学、政治学和社会学等专业，热门专业有农业科学、经济学、表现艺术、应用数学和管理学等专业。在 2021 年福布斯美国最具价值大学排名中，加利福尼亚大学戴维斯校区排名第 20 位，是世界农业与环境科学研究和教育中心。2013—2015 年，加利福尼亚大学戴维斯分校的农林学院占据世界第一。与食品科学相关的科系包括营养学、食品科学与技术。

4）伦敦国王学院（King's College London）（http://www.kcl.ac.uk/index.aspx）

伦敦国王学院是伦敦大学联盟的创校学院之一，位于伦敦泰晤士河畔精华地带，是英国最具生命的多学科、以研究见长的大学之一，其在研究领域的领导地位和世界范围内的极高声誉吸引了来自世界各地的学生在此进修。同时，国王学院也是英国“金三角名校”的一员。国王学院的九大院系包括：人文与艺术学院、生命科学与医学学院、牙科、口腔和颅面科学学院、潘迪生法学院、国王商学院、佛罗伦萨南丁格尔护理与助产学院、自然科学与数学学院、心理学与神经科学学院、社会科学与公共政策学院。截至 2020 年，学校共培养出 16 位诺贝尔奖得主。在生命科学与医学学院开设了两个营养学相关专业，即营养学和饮食营养学。

5）萨里大学（University of Surrey）（http://www.surrey.ac.uk/）

萨里大学位于英格兰东南的萨里郡吉尔福德。该校成立于 1891 年，其前身为伦敦的巴特西理工学院，2022 年 QS 世界大学排名第 272 名。作为英国学术机构的最高荣誉，萨里大学获得了 4 次英国女王周年奖章,其中包括 2017 年食品和营养学研究。该校设有 4 个学院，分别为艺术与人文科学学院，商业、经济与法学院，工程与物理科学学院，健康与医学院。该校开设食品管理专业，其宗旨是培养既懂管理、又懂食品加工业专业知识的综合性人才。课程设置侧重管理知识，将挂历技能和食品产业有机地结合了起来。

6）诺丁汉大学（The University of Nottingham）（http://www.nottingham.ac.uk/）

英国诺丁汉大学建于 1881 年，是英国著名的重点大学，位于英格兰的历史古城诺丁汉市中心附近，是欧洲各国公认并推崇的高等教育学府。诺丁汉大学设有 64 个学系，分属于 5 个学院：工学院、理学院、艺术学院、医学院、社科学院。诺丁汉大学的药学被誉为全世界最顶尖的药学院之一，药剂药理学名列 2021 年 QS 世界大学学科排名第 5 名。

7）雷丁大学（University of Reading）（http://www.reading.ac.uk/）

雷丁大学始建于 1892 年，学校最早是牛津大学创办的分校，于 1926 年得到皇家授权。雷丁大学是英国著名的红砖大学，由于其各方面的成就，如今已成为一所集研究和教学一体的综合大学，位于英国一流大学之列。雷丁大学特别重视科研与研究生教育，学校由 4 个学院组成，包括艺术、人文和社会科学学院、理学院、生命科学学院和亨利商学院。设有 43 个系，五大学科。食品科学方向开设的课程比较全，有食品经济学和市场营销学、食品科学、食品技术及质量管理 3 个授课式硕士专业。

8）利兹大学（University of Leeds）（http://www.leeds.ac.uk/）

利兹大学位于英国第二大金融城市利兹市市中心，是英国规模最大的大学之一，始建于 1831 年，世界百强名校，与曼彻斯特大学等大学并称为英国“红砖大学”。英国常春藤名校联盟“罗素大学集团”的创始成员，与宾夕法尼亚州立大学等同为世界大学联盟成员。

利兹大学与同样位于约克地区的约克大学、雪菲尔大学共同组成“白玫瑰大学联盟”，共同合作在高科技领域的研发工作，联盟整体的研究经费与规模可与牛津大学、剑桥大学相互辉映。利兹大学有 41 个科系，提供 700 多项本科学位课程，研究生学位课程也达 330 多项。2020/2021 年度，利兹大学在 QS 世界大学排名、莱顿世界大学排名和 CWUR 世界大学排名中位列世界百强。大学的研究生院实力雄厚，课程涵盖了文学、社会科学、经济、商务管理、化学、电子通信、信息科技、教育、土木工程、建筑、电子、机械、食品、法律、医学保健、纺织、交通、生物科技、地球科学等，是提供课程最多的英国大学之一。其食品科学在全英首屈一指，具体课程涵盖广泛，包括食品科学、食品生物技术学、食品质量保证和食品与营养学。

9）瓦赫宁根大学与研究中心（Wageningen University and Research Center）（http://www.wageningenur.nl/en/wageningen-university.htm）

瓦赫宁根大学与研究中心（WUR）创建于 1918 年。近几十年来，该校在植物、动物、环境、农业技术、食品和社会科学等领域已发展成为世界上重要的教学和科研中心。该校历史悠久，是荷兰农业研究方面实力最强的大学，在环境科学、食品科学及工程以及生命科学领域是世界顶尖的教育和研究中心。该大学在荷兰高等教育指南上高居榜首，在生命科学领域是欧洲大学的领头羊，也是唯一一所归荷兰农业自然和食品质量部直接拨款的大学。WUR 以“开发自然潜力，提高生活质量”为使命，在“健康食物和生存环境”领域和全球政府机构以及私有研发部门有着紧密的合作关系。WUR 将基础研究、应用研究、教育以及生产实践综合成一体，进行合作创新。同时，采用自然科学与社会科学各学科交叉的方法论来解决现实中面临的复杂问题，如生产供应和食物链问题等。

10）奥塔哥大学（University of Otago）（http://www.otago.ac.nz/）

奥塔哥大学位于新西兰南岛奥塔哥省首府达尼丁市，成立于 1869 年，是新西兰第一所大学。该校有商学、健康科学、人文科学、理学 4 所学院。其研究课题宽泛，学术强项包括生物科学、心理学、人类学、历史和艺术史等。其食品科学的学科组成包括：食品、食品组成、食品质量以及消费者需求。其以培养创新型、具有挑战性、丰富多彩的食品行业优秀职业人才为培养目标。